SOIL MECHANICS LABORATORY MANUAL

Seventh Edition

Braja M. Das
Henderson, Nevada

New York   Oxford
OXFORD UNIVERSITY PRESS
2009

Oxford University Press, Inc., publishes works that further Oxford University's
objective of excellence in research, scholarship, and education.

Oxford New York
Auckland   Cape Town   Dar es Salaam   Hong Kong   Karachi
Kuala Lumpur   Madrid   Melbourne   Mexico City   Nairobi
New Delhi   Shanghai   Taipei   Toronto

With offices in
Argentina Austria Brazil Chile Czech Republic France Greece
Guatemala Hungary Italy Japan Poland Portugal Singapore
South Korea Switzerland Thailand Turkey Ukraine Vietnam

Copyright © 2009 by Oxford University Press, Inc.

Published by Oxford University Press, Inc.
198 Madison Avenue, New York, New York 10016
http://www.oup.com

Oxford is a registered trademark of Oxford University Press

Library of Congress Cataloging-in-Publication Data

Library of Congress Cataloging-in-Publication Data
Das, Braja M., 1941-
    Soil mechanics: laboratory manual / Braja M. Das.– 7th ed.
        p. cm.
    Includes bibliographical references.
    ISBN-13: 978-0-19-536759-1 (pbk.: alk. paper)   1. Soil mechanics–Laboratory
manuals. I. Title.
    TA710.5.D28 2009
    624.1'5136–dc22
                                                    2008012579

Printing number: 9 8 7 6 5 4 3 2 1

Printed in the United States of America
on acid-free paper

CONTENTS

PREFACE

Since the early 1940s the study of soil mechanics has made great progress all over the world. A course in soil mechanics is generally required for undergraduate students in most four-year civil engineering and civil engineering technology programs. The course constitutes classroom lectures along with a laboratory component. The purpose of the laboratory component is to familiarize students with the properties of soils and their behavior under stress by conducting hands-on experiments. This laboratory manual is exclusively prepared for that purpose.

In the United States as well as many other countries around the world, the standards set up for laboratory testing by the American Society for Testing and Materials are used as a guide. The test procedures used in this manual are close to those given by ASTM (2007) with some minor deviations, since the book is intended for classroom use as an initial experience for the students. The procedures and equipment described in this manual are fairly common. For a few tests, such as permeability, direct shear, and unconfined compression, the existing equipment in a given laboratory may differ slightly. In those cases it is necessary that the instructor familiarize students with the operation of the equipment. Triaxial test assemblies are costly, and the equipment varies widely. For that reason only general guidelines for triaxial tests are presented.

For each laboratory test procedure described, sample calculation(s) and graph(s) are included. Also, blank tables and graph paper for each test are provided at the end of the manual for student use in the laboratory and in preparing the final report.

The photographs in this edition are new. The line drawings have been redrawn or revised. Additional discussion at the end of most of the chapters has been provided. Semi-log and linear graph paper have been added in Appendix E.

I would like to thank my wife, Janice F. Das, who apparently possesses endless energy and enthusiasm. She has taken all of the photographs and revised line drawings for this edition. She is also responsible for many of the improvements made in the manual.

Braja M. Das

To our granddaughter, Elizabeth Madison

1
Laboratory Test and Preparation of Report

1.1 Introduction

Proper laboratory testing of soils to determine their physical properties is an integral part in the design and construction of structural foundations, the placement and improvement of soil properties, and the specifications and quality control of soil compaction works. It must be kept in mind that natural soil deposits often exhibit a high degree of nonhomogeneity. The physical properties of a soil deposit can change to a great extent even within a few hundred feet. The fundamental theoretical and empirical equations that are developed in soil mechanics can be properly used in practice if, and only if, the physical parameters used in those equations are properly evaluated in the laboratory. So, learning to perform laboratory tests of soils plays an important role in the geotechnical engineering profession. This text has been prepared exclusively for hands-on classroom use by undergraduate civil engineering and civil engineering technology students taking the introductory soil mechanics (geotechnical engineering) course.

In the United States most laboratories conducting tests on soils for engineering purposes follow the procedures outlined by the American Society for Testing and Materials (ASTM). The procedures and equipment for soil tests may vary slightly from laboratory to laboratory, but the basic concepts remain the same. The test procedures described in this manual may not be exactly the same as specified by ASTM. However, for the students it is beneficial to know the standard test designations and to compare them with the laboratory work they have performed. For this reason, some selected ASTM standard test designations are given in Table 1.1.

Table 1.1. Some Important ASTM Standard Test Designations*

ASTM Title	ASTM Number	Chapter and Comments in This Manual
Laboratory Determination of Water (Moisture) Content of Soil and Rock by Mass	D-2216	2
Specific Gravity of Soil Solids by Water Pycnometer	D-854	3
Particle-Size Analysis of Soils	D-422	4, 5
Liquid Limit, Plastic Limit, and Plasticity Index of Soils	D-4318	6, 7
Standard Test Method for Shrinkage Factors of Soils by the Mercury Method	D-427	8
Standard Test Method for Shrinkage Factors of Soils by the Wax Method	D-4943	Alternative to D-427 (not discussed in this text)
Standard Practice for Classification of Soils and Soil–Aggregate Mixtures for Highway Construction Purposes	D-3282	9 (AASHTO classification system)
Standard Practice for Classification of Soils for Engineering Purposes (Unified Soil Classification System)	D-2484	9
Standard Test Method for Permeability of Granular Soils (Constant Head)	D-2434	10
Standard Test Method for Measurement of Hydraulic Conductivity of Saturated Porous Materials Using a Flexible-Wall Permeameter	D-5084	See Section 11.6
Standard Test Method for Laboratory Compaction Characteristics of Soil Using Standard Effort [12,400 ft · lb/ft^3 (600 kN · m/m^3)]	D-698	12
Standard Test Method for Laboratory Compaction Characteristics of Soil Using Modified Effort [56,000 ft · lb/ft^3 (2700 kN · m/m^3)]	D-1557	13
Standard Test Method for Density and Unit Weight of Soil in Place by the Sand-Cone Method	D-1556	14

Table 1.1. (Continued)

ASTM Title	ASTM Number	Chapter and Comments in This Manual
Standard Test for Density of Soil and Soil–Aggregate in Place by Nuclear Methods (Shallow Depth)	D-2922	Alternative to D-1556 (see Section 14.6)
Standard Test Method for Density and Unit Weight of Soil in Place by the Rubber Balloon Method	D-2167	Alternative to D-1556 (see Section 14.6)
Standard Test Method for Direct Shear Test of Soil under Consolidated Drained Condition	D-3084	15
Standard Test Method for Unconfined Compression Strength of Cohesive Soil	D-2166	16
Standard Test Method for One Dimensional Consolidation Properties of Soils Using Incremental Loading	D-2435	17
Standard Test Method for Unconsolidated Undrained Triaxial Compression Test on Cohesive Soils	D-2850	18
Standard Test Method for Consolidated Undrained Triaxial Compression Test on Cohesive Soils	D-4767	18

*Based on American Society for Testing and Materials (2007).

1.2 Use of Equipment

Laboratory equipment is never cheap, but the cost may vary widely. For accuracy of the experimental results, the equipment should be properly maintained. The calibration of certain equipment, such as balances and proving rings, should be checked periodically. It is also essential that all equipment be clean—both before and after use. More accurate results will be obtained when the equipment being used is clean, so always maintain the equipment as if it were your own.

1.3 Safety

There is always a possibility that an accident may occur while performing a test in the laboratory and/or in the field. Proper care must be taken to prevent such accidents from occurring. Hazardous material, such as mercury (see Chapter 8), must be handled carefully, and spills should be avoided.

1.4 Data Recording

In any experiment it is always good practice to record all data in the proper table immediately after they have been taken. Scribbles on scratch paper may later be illegible or even misplaced, which may result in having to conduct the experiment again or in obtaining inaccurate results.

1.5 Report Preparation

In the classroom laboratory, most experiments described herein will probably be conducted in small groups. However, the laboratory report should be written by each student individually. This is one way that students can improve their technical writing skills. Each report should contain:

1. Cover page—The cover page should include the title of the experiment, name, group number, and date on which the experiment was performed.
2. Following the cover page, the following items should be included in the body of the report:
   a. Purpose of experiment
   b. Equipment used
   c. Schematic diagram of main equipment used
   d. Brief description of test procedure
3. Results—These should include the data sheet(s), sample calculation(s), and required graph(s). Graphs and tables should be prepared as neatly as possible. *Always* give the units. Graphs should be as large as possible, and they should be labeled properly. When necessary, French curves and a straightedge should be used in preparing graphs. Graph paper is provided in Appendix E. When a computer is used to draw a graph, its authenticity should be properly verified.

1.6 Units

It may be necessary to express the results of laboratory tests in a given system of units. Both English and SI units are used at this time in the United States. Conversion of units may be necessary in preparing the reports. Some selected conversion factors from English to SI units and from SI to English units are given in Appendix B.

2
Determination of Water Content

2.1 Introduction

Most laboratory tests in soil mechanics require determination of the water content. Water content is defined as

$$w = \frac{\text{weight (or mass) of water present in a given soil mass}}{\text{weight (or mass) of dry soil}} \qquad (2.1)$$

Water content is usually expressed in percent.

For better results, the *minimum* size of most soil specimens should be approximately as given in Table 2.1. These values are consistent with ASTM Test Designation D-2216.

2.2 Equipment

1. Moisture can(s)—Moisture cans are available in various sizes [for example, 2 in. (50.8 mm) in diameter and $7/8$ in. (22.2 mm) high; 3.5 in. (88.9 mm) in diameter and 2 in. (50.8 mm) high].
2. Oven with temperature control—For drying, the oven temperature is generally kept at $110 \pm 5°C$. A higher temperature should be avoided to prevent the burning of organic matter in the soil.
3. Balance—The readability of the balance to be used is given in Table 2.2 (ASTM, 2007). Figure 2.1 shows some moisture cans and a balance having a readability of 0.01 g.

Table 2.1. Minimum Size of Moist Soil Samples to Determine Water Content

Maximum Particle Size in Soil (mm)	U.S. Sieve No.	Minimum Mass of Soil Sample (g)
0.425	40	20
2.0	10	50
4.75	4	100
9.5	3/8 in.	500
19.0	3/4 in.	2500
37.5	1.5 in.	10,000
75.0	3.0 in.	50,000

Table 2.2. Required Readability of Balance

Maximum Particle Size in Soil (mm)	Readability of Balance (g)
0.425	0.01
2.0	0.01
4.75	0.1
9.5	0.1
19.0	1
37.5	10
75.0	10

2.3 Procedure

1. Determine the mass (g) of the empty moisture can plus its cap, M_1, and also record the number.
2. Place a sample of representative moist soil in the can. Close the can with its cap to avoid loss of moisture.
3. Determine the combined mass (g) of the closed can and moist soil, M_2.

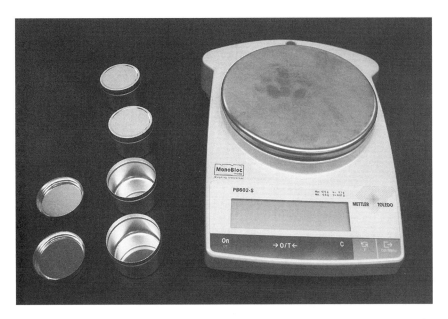

Figure 2.1. Moisture cans and balance.

4. Remove the cap from the top of the can and place it on the bottom (of the can).
5. Put the can (Step 4) in the oven to dry the soil to a constant weight. In most cases 24 hours of drying is enough.
6. Determine the combined mass (g) of the dry soil sample plus the can and its cap, M_3.

2.4 Calculations

1. Calculate the mass of moisture, $M_2 - M_3$
2. Calculate the mass of dry soil, $M_3 - M_1$
3. Calculate the water content,

$$w\,(\%) = \frac{M_2 - M_3}{M_3 - M_1} \times 100 \tag{2.2}$$

Report the water content to the nearest 1% or 0.1%, as appropriate, based on the size of the specimen.

    A sample calculation of water content is given in Table 2.3.

2.5 General Comments

1. Most natural soils that are sandy and gravelly in nature may have water contents of up to about 15–20%. In natural fine-grained (silty or clayey) soils, water contents of up to about 50–80% can be found. However, peat and highly organic soils with water contents of up to about 500% are not uncommon. Typical values of water content for various types of natural soils in a saturated state are shown in Table 2.4.

Table 2.3. Determination of Water Content

Description of soil _____ *Brown silty clay* _____ Sample no. ___ *4* ___

Location _____

Tested by _____ Date _____

Item	Test No.		
	1	2	3
Can no.	42	31	54
Mass of can, M_1 (g)	17.31	18.92	16.07
Mass of can + wet soil, M_2 (g)	43.52	52.19	39.43
Mass of can + dry soil, M_3 (g)	39.86	47.61	36.13
Mass of moisture, $M_2 - M_3$ (g)	3.66	4.58	3.30
Mass of dry soil, $M_3 - M_1$ (g)	22.55	28.69	20.06
Water content, w (%) $= \dfrac{M_2 - M_3}{M_3 - M_1} \times 100$	16.2	16.0	16.5

Average water content w ___ *16.2* ___ %

Table 2.4. Typical Values of Water Content
in a Saturated State

Soil	Natural Water Content in a Saturated State (%)
Loose uniform sand	25–30
Dense uniform sand	12–16
Loose angular-grained silty sand	25
Dense angular-grained silty sand	15
Stiff clay	20
Soft clay	30–50
Soft organic clay	80–130
Glacial till	10

2. Some organic soils may decompose during oven drying at $110 \pm 5°C$. This oven drying temperature may be too high for soils containing gypsum, as this material dehydrates slowly. For such soils a drying temperature of $60°C$ is more appropriate.
3. Cooling the dry soil after oven drying (Step 5) in a desiccator is recommended. It prevents absorption of moisture from the atmosphere.

3

Specific Gravity of Soil Solids

3.1 Introduction

The specific gravity of a given material is defined as the ratio of the density of a given volume of the material to the density of an equal volume of distilled water. In soil mechanics, the specific gravity of soil solids (which is often referred to as the specific gravity of soil) is an important parameter for calculating the weight–volume relationship. Thus specific gravity G_s is defined as

$$G_s = \frac{\text{density of soil solids only}}{\text{density of water}}$$

or

$$G_s = \frac{M_s/V_s}{\rho_w} = \frac{M_s}{V_s \rho_w} \tag{3.1}$$

where M_s = mass of soil solids (g)
V_s = volume of soil solids (cm^3)
ρ_w = density of water (g/cm^3)

Most soils found in nature are combinations of various types of minerals. The ranges of the values of G_s for common minerals found in soil are given in Table 3.1. The general ranges of the values of G_s for various soils are given in Table 3.2. The procedure for determining the specific gravity G_s described here is applicable for soils composed of particles *smaller than* 4.75 mm in size (No. 4 U.S. sieve).

Table 3.1. General Ranges of G_s for
Common Minerals

Mineral	Range of G_s
Quartz	2.65
Kaolinite	2.6
Illite	2.8
Montmorillonite	2.65–2.80
Halloysite	2.0–2.55
Potassium feldspar	2.57
Sodium and calcium feldspar	2.62–2.76
Chlorite	2.6–2.9
Biotite	2.8–3.2
Muscovite	2.76–3.1
Hornblende	3.0–3.47
Limonite	3.6–4.0
Olivine	3.27–3.7

Table 3.2. General Ranges of G_s for
Various Soils

Soil Type	Range of G_s
Sand	2.63–2.67
Silts	2.65–2.7
Clay and silty clay	2.67–2.9
Organic soil	Less than 2

3.2 Equipment

1. Volumetric flask (500 ml)
2. Thermometer graduated in 0.5°C division scale

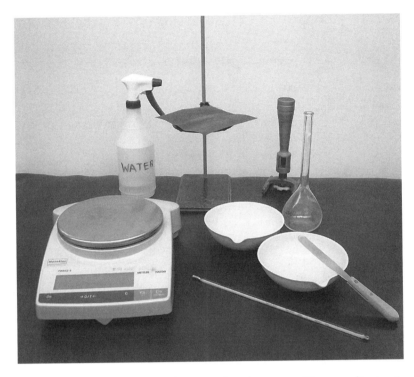

Figure 3.1. Equipment for conducting specific gravity test.

3. Balance sensitive up to 0.01 g
4. Distilled water
5. Bunsen burner and stand (and/or vacuum pump or aspirator)
6. Evaporating dishes
7. Spatula
8. Plastic squeeze bottle
9. Drying oven

The equipment for this experiment is shown in Fig. 3.1.

3.3 Procedure

1. Clean the volumetric flask well and dry it.
2. Carefully fill the flask with de-aired, distilled water up to the 500-ml mark (bottom of the meniscus should be at the 500-ml mark).
3. Determine the mass of the flask and the water filled to the 500-ml mark, M_1.
4. Insert the thermometer into the flask with the water and determine the temperature of the water, $T = T_1°C$.
5. Put the air-dried soil into an evaporating dish. Table 3.3 gives the approximate mass of dry soil to be used for the test (ASTM test designation D-854).

Table 3.3.   Approximate Mass of Dry Soil to Be Used for Test

Soil Type*	General Description	Dry Mass of Specimen (g)
SP, SP–SM	Poorly graded sand; poorly graded sand with silt	100 ± 10
SP–SC, SM, SC	Poorly graded sand with clay; silty sand; clayey sand	75 ± 10
Silt and clay	—	50 ± 10

*Unified soil classification; see Chapter 9.

6. If the soil is cohesive, add water (de-aired and distilled) to the soil and mix it to the form of a smooth paste. Keep it soaked in the evaporating dish for about one-half to one hour. (*Note:* This step is not necessary for granular, i.e., noncohesive soils.)
7. Transfer the soil (if granular) or the soil paste (if cohesive) into the volumetric flask.
8. Add distilled water to the volumetric flask containing the soil (or the soil paste) to make it about two-thirds full.
9. Remove the air from the soil–water mixture. This can be done by either:
   a. Gently boiling the flask containing the soil–water mixture for about 15–20 minutes. Accompany the boiling with continuous agitation of the flask. (If too much heat is applied, the soil may boil over.) or
   b. Applying vacuum by a vacuum pump or aspirator until all of the entrapped air is out.

   This is an *extremely important* step. Most errors in the results of this test are due to *entrapped air that is not removed*.
10. Bring the temperature of the soil–water mixture in the volumetric flask down to room temperature, i.e., $T_1°C$; see Step 4. (This temperature of the water is room temperature.)
11. Add de-aired, distilled water to the volumetric flask until the bottom of the meniscus touches the 500-ml mark. Also dry the outside of the flask and the inside of the neck above the meniscus.
12. Determine the combined mass of bottle plus soil plus water, M_2.
13. Just as a precaution, check the temperature of soil and water in the flask to see whether or not it is $T_1 \pm 1°C$.
14. Pour the soil and water into an evaporating dish. Use a plastic squeeze bottle and wash the inside of the flask. Make sure that no soil is left inside.
15. Put the evaporating dish in an oven to dry to a constant weight.
16. Determine the mass of the dry soil in the evaporating dish, M_s.

3.4 Calculations

Calculate the specific gravity,

$$G_s = \frac{\text{mass of soil } M_s}{\text{mass of equal volume of water } M_w}$$

where the mass of equal volume of water is,

$$M_w = (M_1 + M_s) - M_2$$

So,

$$G_{s\,(\text{at } T_1 °C)} = \frac{M_s}{M_w} \tag{3.2}$$

Specific gravity is generally reported at a temperature corresponding to 20°C. So,

$$G_{s\,(\text{at } 20°C)} = G_{s\,(\text{at } T_1 °C)} \left[\frac{\rho_{w\,(\text{at } T_1 °C)}}{\rho_{w\,(\text{at } 20°C)}} \right]$$

$$= G_{s\,(\text{at } T_1 °C)}\, A \tag{3.3}$$

where

$$A = \frac{\rho_{w\,(\text{at } T_1 °C)}}{\rho_{w\,(\text{at } 20°C)}} \tag{3.4}$$

and ρ_w is the density of water. The values of A are given in Table 3.4.

Table 3.4.   Values of A [Eq. (3.4)]

Temperature (T_1°C)	A	Temperature (T_1°C)	A
16	1.0007	24	0.9991
17	1.0006	25	0.9988
18	1.0004	26	0.9986
19	1.0002	27	0.9983
20	1.0000	28	0.9980
21	0.9998	29	0.9977
22	0.9996	30	0.9974
23	0.9993		

Table 3.5. Specific Gravity of Soil Solids

Description of soil _____ *Light brown sandy silt* _____ Sample no. ___ *23* ___

Volume of flask at 20°C _*500*_ ml Temperature of test _*23*_ °C A _*0.9993*_ (Table 3.4)

Location _____

Tested by _____ Date _____

Item	Test No.		
	1	**2**	**3**
Volumetric flask no.	6	8	9
Mass of flask + water filled to mark, M_1 (g)	666.0	674.0	652.0
Mass of flask + soil + water filled to mark, M_2 (g)	722.0	738.3	709.93
Mass of dry soil, M_s (g)	99.0	103.0	92.0
Mass of equal volume of water and soil solids, M_w (g) $= (M_1 + M_s) - M_2$	37.0	38.7	34.07
$G_{s\,(\text{at } T_1°C)} = M_s/M_W$	2.68	2.66	2.70
$G_{s\,(\text{at } T_1°C)} = G_{s\,(\text{at } T_1°C)} \times A$	2.68	2.66	2.70

Average $G_s = \dfrac{2.68 + 2.66 + 2.70}{3} = 2.68$

At least three specific gravity tests should be conducted. For correct results, these values should not vary by more than 2–3%. A sample calculation for specific gravity is shown in Table 3.5.

4
Sieve Analysis

4.1 Introduction

In order to classify a soil for engineering purposes, one needs to know the distribution of the grain sizes in a given soil mass. Sieve analysis is a method used to determine the grain size distribution of soils. Sieves are made of woven wires with square openings. Note that as the sieve number increases, the size of the openings decreases. Table 4.1 lists the U.S. standard sieve numbers with their corresponding opening sizes. For all practical purposes, the No. 200 sieve is the sieve with the smallest opening that should be used for the test. The sieves that are most commonly used for soil tests have a diameter of 8 in. (203 mm). A stack of sieves is shown in Fig. 4.1.

The method of sieve analysis described here is applicable for soils that are *mostly granular, with some or no fines*. Sieve analysis does not provide information about the shape of the particles.

4.2 Equipment

1. Sieves, a bottom pan, and a cover. (*Note:* Sieve numbers 4, 10, 20, 40, 60, 140, and 200 are generally used for most standard sieve analysis work.)
2. A balance sensitive to 0.1 g
3. Mortar and rubber-tipped pestle
4. Oven
5. Mechanical sieve shaker

4.3 Procedure

1. Collect a *representative* oven-dry soil sample. Samples with the largest particles being of the size of No. 4 sieve openings (4.75 mm) should weigh about 500 g.

Table 4.1. U.S. Sieve Sizes

Sieve No.	Opening (mm)	Sieve No.	Opening (mm)
4	4.75	35	0.500
5	4.00	40	0.425
6	3.35	45	0.355
7	2.80	50	0.300
8	2.36	60	0.250
10	2.00	70	0.212
12	1.70	80	0.180
14	1.40	100	0.150
16	1.18	120	0.125
18	1.00	140	0.106
20	0.85	200	0.075
25	0.71	270	0.053
30	0.60	400	0.038

Figure 4.1. Stack of sieves with a pan at the bottom and a cover at the top.

For soils with the largest particles of a size greater than 4.75 mm, larger weights are needed.

2. Break the soil sample into individual particles using a mortar and a rubber-tipped pestle. (*Note:* The idea is to break up the soil into individual particles, not to break the particles themselves.)

3. Determine the mass M of the sample accurately to 0.1 g.

4. Prepare a stack of sieves. A sieve with larger openings is placed above a sieve with smaller openings. The sieve at the bottom should be No. 200. A bottom pan should be placed under the No. 200 sieve. As mentioned before, the sieves that are generally used in a stack are Nos. 4, 10, 20, 40, 60, 140, and 200; however, more sieves can be placed in between.

5. Pour the soil prepared in Step 2 into the stack of sieves from the top.

6. Place the cover on the top of the stack of sieves.

7. Run the stack of sieves through a sieve shaker for about 10–15 minutes (Fig. 4.2).

8. Stop the sieve shaker and remove the stack of sieves.

9. Weigh the amount of soil retained on each sieve and in the bottom pan.

10. If a *considerable* amount of soil with silty and clayey fractions is retained on the No. 200 sieve, it has to be washed. Washing is done by taking the No. 200 sieve with the soil retained on it and pouring water through the sieve from a tap in the laboratory (Fig. 4.3).

Figure 4.2. Stack of sieves in a sieve shaker.

Figure 4.3. Washing of soil retained on No. 200 sieve.

11. When the water passing through the sieve is clean, stop the flow of water. Transfer the soil retained on the sieve at the end of washing to a porcelain evaporating dish by back washing (Fig. 4.4). Put it in the oven to dry to a constant weight. (*Note:* This step is not necessary if the amount of soil retained on the No. 200 sieve is small.)

Determine the mass of the dry soil retained on the No. 200 sieve. The difference between this mass and that retained on the No. 200 sieve determined in Step 9 is the mass of soil that has washed through.

4.4 Calculations

1. Calculate the percent of soil retained on the *n*th sieve (counting from the top),

$$\frac{\text{mass retained } M_n}{\text{total mass } M \text{ (Step 3)}} \times 100 = R_n \tag{4.1}$$

2. Calculate the cumulative percent of soil retained on the *n*th sieve,

$$\sum_{i=1}^{i=n} R_n \tag{4.2}$$

Figure 4.4. Back washing to transfer soil retained on No. 200 sieve to an evaporating dish.

3. Calculate the cumulative percent passing through the *n*th sieve,

$$\text{percent finer} = 100 - \sum_{i=1}^{i=n} R_n \qquad (4.3)$$

Note: If soil retained on the No. 200 sieve is washed, the dry unit weight determined after washing (Step 11) should be used to calculate the percent finer (than No. 200 sieve). The weight lost due to washing should be added to the weight of the soil retained on the pan.

A sample calculation of sieve analysis is shown in Table 4.2.

4.5 Graphs

The grain-size distribution obtained from the sieve analysis is plotted on semilogarithmic graph paper with the grain size on a log scale and percent finer on a natural scale. Figure 4.5 is a grain-size distribution plot for the calculations shown in Table 4.2. The grain-size distribution plot helps to estimate the percent finer than a given sieve size that might not have been used during the test.

Table 4.2. Sieve Analysis

Description of soil _____ *Sand with some fines* _____ Sample no. _____2_____

Mass of oven-dry specimen M _500_ g

Location _____

Tested by _____ Date _____

Sieve No.	Sieve Opening (mm)	Mass of Soil Retained on Each Sieve M_n (g)	Percent of Mass Retained on Each Sieve R_n	Cumulative Percent Retained $\sum R_n$	Percent Finer $100 - \sum R_n$
4	4.750	0	0	0	100.0
10	2.000	40.2	8.0	8.0	92.0
20	0.850	84.6	16.9	24.9	75.1
30	0.600	50.2	10.0	34.9	65.1
40	0.425	40.0	8.0	42.9	57.1
60	0.250	106.4	21.3	64.2	35.8
140	0.106	108.8	21.8	86.0	14.0
200	0.075	59.4	11.9	97.9	2.1
Pan	____	8.7			

$\sum$ _498.3_ $= M_1$

Mass loss during sieve analysis: $\dfrac{M - M_1}{M} \times 100 =$ _0.34_ % (OK if less than 2%)

4.6 Other Calculations

1. Determine D_{10}, D_{30}, and D_{60} (from Fig. 4.5), which are the diameters corresponding to percents finer of 10%, 30%, and 60%, respectively.
2. Calculate the uniformity coefficient C_u and the coefficient of gradation C_c using the following equations:

$$C_u = \frac{D_{60}}{D_{10}} \tag{4.4}$$

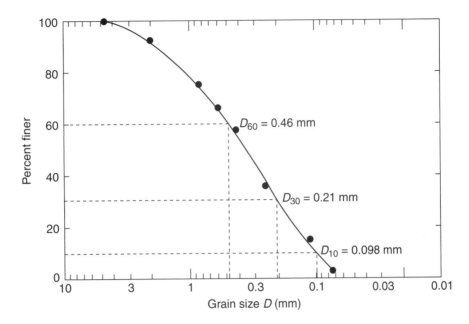

Figure 4.5. Plot of percent finer vs. grain-size from calculations shown in Table 4.2.

$$C_c = \frac{D_{30}^2}{D_{60} \times D_{10}} \tag{4.5}$$

As an example, from Fig. 4.5, $D_{60} = 0.46$ mm, $D_{30} = 0.21$ mm, and $D_{10} = 0.098$ mm. So,

$$C_u = \frac{0.46}{0.098} = 4.69$$

and

$$C_c = \frac{(0.21)^2}{0.46 \times 0.098} = 0.98$$

4.7 General Comments

The diameter D_{10} is generally referred to as *effective size*. The effective size is used for several empirical correlations, such as the *coefficient of permeability*. The *coefficient of gradation* C_u is a parameter that indicates the range of distribution of the grain sizes in a given soil specimen. If C_u is relatively large, it indicates a well-graded soil. If C_u is nearly equal to 1, it means that the soil grains are of approximately equal size, and the soil may be referred to as a poorly graded soil.

Figure 4.6 shows the general nature of the grain-size distribution curves for a well-graded and a poorly graded soil. In some instances a soil may have a combination of two or more uniformly graded fractions, and this soil is referred to as gap graded. The grain-size distribution curve for a gap-graded soil is also shown in Fig. 4.6.

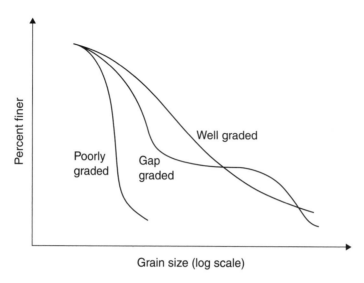

Figure 4.6. General nature of grain-size distribution of well-graded, poorly graded, and gap-graded soil.

The parameter C_c is also referred to as the *coefficient of curvature.* For sand, if C_u is greater than 6 and C_c between 1 and 3, it is considered well graded. However, for a gravel to be well graded, C_u should be greater than 4 and C_c must be between 1 and 3.

The D_{15} and D_{85} sizes are used for the design of filters. The D_{50} size is used for correlation of the liquefaction potential of saturated granular soil during earthquakes.

5
Hydrometer Analysis

5.1 Introduction

Hydrometer analysis is the procedure generally adopted for determining the particle-size distribution in a soil for the fraction that is finer than No. 200 sieve size (0.075 mm). The lower limit of the particle size determined by this procedure is about 0.001 mm.

In hydrometer analysis a soil specimen is dispersed in water. In a dispersed state in the water, the soil particles will settle individually. It is assumed that the soil particles are spheres, and the velocity of the particles can be given by Stoke's law,

$$v = \frac{\gamma_s - \gamma_w}{18\eta} D^2 \qquad (5.1)$$

where $v =$ velocity (cm/s)
$\gamma_s =$ specific weight of soil solids (g/cm^3)
$\gamma_w =$ unit weight of water (g/cm^3)
$\eta =$ viscosity of water (g · s/cm^2)
$D =$ diameter of soil particle

In the test procedure described here, the ASTM 152 H type hydrometer will be used (see Fig. 5.1). If a hydrometer is suspended in water in which soil is dispersed (Fig. 5.2), it will measure the specific gravity of the soil–water suspension at a depth L. The depth L is called *effective depth*. So at a time t minutes from the beginning of the test, the soil particles that settle beyond the zone of measurement (i.e., beyond the effective depth L) will have a diameter given by

$$\frac{L \,(\text{cm})}{t \,(\text{min}) \times 60} = \frac{\gamma_s - \gamma_w \,(\text{g/cm}^3)}{18\eta \,(\text{g} \cdot \text{s/cm}^2)} \left[\frac{D \,(\text{mm})}{10} \right]^2$$

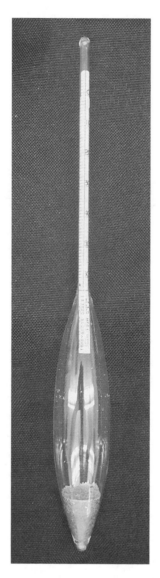

Figure 5.1. ASTM 152 H hydrometer.

Then

$$D \text{ (mm)} = \frac{10}{\sqrt{60}} \sqrt{\frac{18\eta}{\gamma_s - \gamma_w}} \sqrt{\frac{L}{t}} = A \sqrt{\frac{L \text{ (cm)}}{t \text{ (min)}}} \qquad (5.2)$$

where

$$A = \sqrt{\frac{1800\eta}{60 \, (\gamma_s - \gamma_w)}} = \sqrt{\frac{30\eta}{\gamma_s - \gamma_w}} \qquad (5.3)$$

From Fig. 5.2 it can be seen that, based on the hydrometer reading (which increases from zero to 60 in the ASTM 152-H type hydrometer), the value of L will change. The magnitude

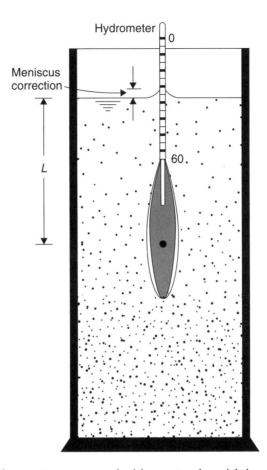

Figure 5.2. Hydrometer suspended in water in which soil is dispersed.

of L can be given as

$$L = L_1 + \frac{1}{2}\left(L_2 - \frac{V_B}{A_C}\right) \tag{5.4}$$

where L_1 = distance between top of hydrometer bulb and mark for a hydrometer reading. For a hydrometer reading of zero, $L_1 = 10.5$ cm. Also, for a hydrometer reading of 50 g/l, $L_1 = 2.3$ cm. Thus, in general, for a given hydrometer reading,

$$L_1 \text{ (cm)} = 10.5 - \left(\frac{10.5 - 2.3}{50}\right) \times \text{hydrometer reading}$$

$L_2 = 14$ cm
V_B = volume of hydrometer bulb $= 67.0$ cm^3
A_C = cross-sectional area of hydrometer cylinder $= 27.8$ cm^2

Based on Eq. (5.4), the variations of L with the hydrometer readings are shown in Table 5.1.

Table 5.1. Variations of *L* with Hydrometer* Readings

Hydrometer Reading	*L* (cm)	Hydrometer Reading	*L* (cm)
0	16.3	26	12.0
1	16.1	27	11.9
2	16.0	28	11.7
3	15.8	29	11.5
4	15.6	30	11.4
5	15.5	31	11.2
6	15.3	32	11.1
7	15.2	33	10.9
8	15.0	34	10.7
9	14.8	35	10.6
10	14.7	36	10.4
11	14.5	37	10.2
12	14.3	38	10.1
13	14.2	39	9.9
14	14.0	40	9.7
15	13.8	41	9.6
16	13.7	42	9.4
17	13.5	43	9.2
18	13.3	44	9.1
19	13.2	45	8.9
20	13.0	46	8.8
21	12.9	47	8.6
22	12.7	48	8.4
23	12.5	49	8.3
24	12.4	50	8.1
25	12.2	51	7.9

*ASTM 152-H hydrometer.

For actual calculation purposes we also need to know the values of A given by Eq. (5.3). An example of this calculation is

$$\gamma_s = G_s \gamma_w$$

where G_s is the specific gravity of soil solids. Thus,

$$A = \sqrt{\frac{30\eta}{(G_s - 1)\gamma_w}} \qquad (5.5)$$

For example, if the temperature of the water is 25°C, $\eta = 0.0911 \times 10^{-4}$ (g·s/cm^2), and $G_s = 2.7$,

$$A = \sqrt{\frac{30\left(0.0911 \times 10^{-4}\right)}{(2.7 - 1)(1)}} = 0.0127$$

The variations of A with G_s and the water temperature are shown in Table 5.2.

The ASTM 152-H type hydrometer is calibrated up to a reading of 60 at a temperature of 20°C for soil particles having $G_s = 2.65$. A hydrometer reading of, say, 30 at a given time of a test means that there are 30 g of soil solids ($G_s = 2.65$) in suspension per 1000 cm^3 of soil–water mixture at a temperature of 20°C at a depth where the specific gravity of the soil–water suspension is measured (i.e., L). From this measurement we can determine the percentage of soil still in suspension at time t from the beginning of the test, and all the soil particles will have diameters smaller than D calculated by Eq. (5.2). However, in the actual experimental work, some corrections to the observed hydrometer readings need to be applied. They are as follows:

1. Temperature correction F_T—The actual temperature of the test may not be 20°C. The temperature correction F_T may be approximated as

$$F_T = -4.85 + 0.25T \quad \left(\text{for } T \text{ between 15 and 28°C}\right) \qquad (5.6)$$

   where F_T = temperature correction to observed reading (can be either positive or negative)
   T = temperature of test (°C)

2. Meniscus correction F_m—Generally the upper level of the meniscus is taken as the reading during laboratory work (F_m is always positive).
3. Zero correction F_z—A deflocculating agent is added to the soil–distilled water suspension when performing experiments. This will change the zero reading (F_z can be either positive or negative).

Table 5.2. Variations of A with G_s

G_s	Temperature (°C)						
	17	**18**	**19**	**20**	**21**	**22**	**23**
2.50	0.0149	0.0147	0.0145	0.0143	0.0141	0.0140	0.0138
2.55	0.0146	0.0144	0.0143	0.0141	0.0139	0.0137	0.0136
2.60	0.0144	0.0142	0.1040	0.0139	0.0137	0.0135	0.0134
2.65	0.0142	0.0140	0.0138	0.0137	0.0135	0.0133	0.0132
2.70	0.0140	0.0138	0.1036	0.0134	0.0133	0.0131	0.0130
2.75	0.0138	0.0136	0.0134	0.0133	0.0131	0.0129	0.0128
2.80	0.0136	0.0134	0.0132	0.0131	0.0129	0.0128	0.0126

G_s	Temperature (°C)						
	24	**25**	**26**	**27**	**28**	**29**	**30**
2.50	0.0137	0.0135	0.0133	0.0132	0.0130	0.0129	0.0128
2.55	0.0134	0.0133	0.0131	0.0130	0.0128	0.0127	0.0126
2.60	0.0132	0.0131	0.0129	0.0128	0.0126	0.0125	0.0124
2.65	0.0130	0.0129	0.0127	0.0126	0.0124	0.0123	0.0122
2.70	0.0128	0.0127	0.0125	0.0124	0.0123	0.0121	0.0120
2.75	0.0126	0.0125	0.0124	0.0122	0.0121	0.0120	0.0118
2.80	0.0125	0.0123	0.0122	0.0120	0.0119	0.0118	0.0117

5.2 Equipment

1. ASTM 152-H hydrometer
2. Mixer
3. Two 1000-cm^3 graduated cylinders
4. Thermometer
5. Constant-temperature bath
6. Deflocculating agent
7. Spatula

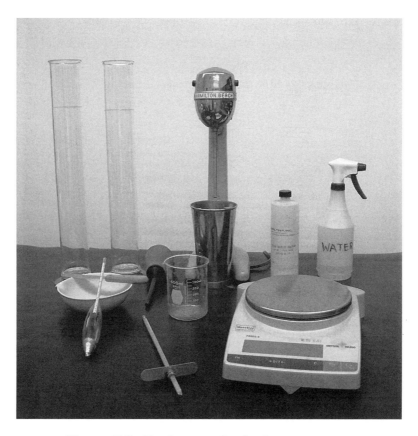

Figure 5.3. Equipment for hydrometer test.

8. Beaker
9. Balance
10. Plastic squeeze bottle
11. Distilled water
12. No. 12 rubber stopper

The equipment necessary (except for the balance and the constant-temperature bath) is shown in Fig. 5.3.

5.3 Procedure

Note: This procedure should be used when more than 90% of the soil is finer than No. 200 sieve.

1. Take 50 g of oven-dry, well pulverized soil in a beaker.
2. Prepare a deflocculating agent. Usually a 4% solution of sodium hexametaphosphate (Calgon) is used. This can be prepared by adding 40 g of Calgon in 1000-cm^3 of distilled water and mixing it thoroughly.

3. Take 125 cm^3 of the mixture prepared in Step 2 and add it to the soil taken in Step 1. This should be allowed to soak for about 8–12 hours.
4. Take a 1000-cm^3 graduated cylinder and add 875 cm^3 of distilled water *plus* 125 cm^3 of deflocculating agent to it. Mix the solution well.
5. Put the cylinder (from Step 4) in a constant-temperature bath. Record the temperature T of the bath (°C).
6. Put the hydrometer in the cylinder (Step 5). Record the reading. (*Note:* The *top of the meniscus* should be read.) This is the zero correction F_z, which can be positive or negative. Also observe the meniscus correction F_m.
7. Using a spatula, mix the soil prepared in Step 3 thoroughly. Pour it into the mixer cup. (*Note:* During this process some soil may stick to the side of the beaker. Using the plastic squeeze bottle filled with distilled water, wash all the remaining soil in the beaker into the mixer cup.)
8. Add distilled water to the cup to make it about two-thirds full. Mix it for about 2 minutes using the mixer.
9. Pour the mix into the second graduated 1000-cm^3 cylinder. Make sure that all of the soil solids are washed out of the mixer cup. Fill the graduated cylinder with distilled water to bring the water level up to the 1000-cm^3 mark.
10. Secure a No. 12 rubber stopper on the top of the cylinder (Step 9). Mix the soil–water well by turning the cylinder upside down several times.
11. Put the cylinder into the constant-temperature bath next to the cylinder described in Step 5. Record the time immediately. This is cumulative time $t = 0$. Insert the hydrometer into the cylinder containing the soil–water suspension.
12. Take hydrometer readings at cumulative times $t = 0.25, 0.5, 1,$ and 2 min. Always read the upper level of the meniscus.
13. Take the hydrometer out after 2 min and put it into the cylinder next to it (Step 5).
14. Hydrometer readings are to be taken at times $t = 4, 8, 15, 30$ minutes, 1, 2, 4, 8, 24, and 48 hours. For each reading, insert the hydrometer into the cylinder containing the soil–water suspension about 30 seconds before the reading is due. After the reading has been taken, remove the hydrometer and put it back into the cylinder next to it (Step 5).

5.4 Calculations

Refer to Table 5.3.

Column 2. Observed hydrometer readings R corresponding to times given in column 1.
Column 3. Corrected hydrometer readings R_{cp} for calculation of percent finer,

$$R_{cp} = R + F_T - F_z \tag{5.7}$$

Column 4. Percent finer $= \dfrac{aR_{cp}}{M_s} \times 100$

Table 5.3. Hydrometer Analysis

Description of soil _____ *Brown silty clay* _____ Sample no. _____

Location _____

G_s _____ 2.75 _____ Hydrometer type _____ *ASTM 152-H* _____

Dry mass of soil M_s _____ 50 _____ g   Temperature of test T _____ 28 _____ °C

Meniscus correction F_m _1_   Zero correction F_z _+7_

Temperature correction F_T _+2.15_ [Eq. (5.6)]

Tested by _____ Date _____

Time (min)	Hydrometer Reading R	R_{cp}	Percent Finer, $\frac{a*R_{cp}}{50} \times 100$	R_{cL}	$L^\dagger$ (cm)	$A^\ddagger$	D (mm)
(1)	(2)	(3)	(4)	(5)	(6)	(7)	(8)
0.25	51	46.15	90.3	52	7.8	0.0121	0.068
0.5	48	43.15	84.4	49	8.3		0.049
1	47	42.15	82.4	48	8.4		0.035
2	46	41.15	80.5	47	8.6		0.025
4	45	40.15	78.5	46	8.8		0.018
8	44	39.15	76.6	45	8.95		0.013
15	43	38.15	74.6	44	9.1		0.009
30	42	37.15	72.7	43	9.25		0.007
60	40	35.15	68.8	41	9.6		0.005
120	38	33.15	64.8	39	9.9		0.0035
240	34	29.15	57.0	35	10.5		0.0025
480	32	27.15	53.1	33	10.9		0.0018
1440	29	24.15	47.23	30	11.35		0.0011
2880	27	22.15	43.3	28	11.65		0.0008

*Table 5.4; †Table 5.1; ‡Table 5.2.

where M_s = dry mass of soil used for hydrometer analysis
a = correction for specific gravity (since hydrometer is calibrated for $G_s = 2.65$),

$$a = \frac{G_s \times 1.65}{(G_s - 1)\,2.65} \quad \text{(see Table 5.4)} \tag{5.8}$$

Column 5. Corrected reading R_{cL} for determination of effective length,

$$R_{cL} = R + F_m \tag{5.9}$$

Column 6. Determine L (effective length) corresponding to values of R_{cL} (column 5) given in Table 5.1.
Column 7. Determine A from Table 5.2.
Column 8. Determine D,

$$D \text{ (mm)} = A\sqrt{\frac{L \text{ (cm)}}{t \text{ (min)}}}$$

5.5 Graphs

Plot a grain-size distribution graph on semilog graph paper with percent finer (column 4, Table 5.3) on a natural scale and D (column 8, Table 5.3) on a log scale. A sample calculation and the corresponding graph are shown in Table 5.3 and Fig. 5.4, respectively.

Table 5.4. Variation of *a* with G_s [Eq. (5.8)]

G_s	a
2.50	1.04
2.55	1.02
2.60	1.01
2.65	1.00
2.70	0.99
2.75	0.98
2.80	0.97

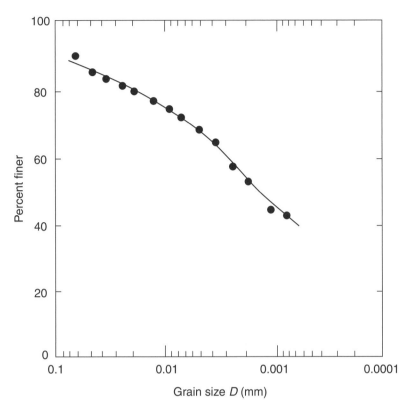

Figure 5.4. Plot of percent finer vs. grain size from results given in Table 5.3.

5.6 Procedure Modification

When a smaller amount (less than about 90%) of soil is finer than No. 200 sieve size, the following modifications to the procedure described need to be applied.

1. Take an oven-dry sample of soil. Determine its weight M_1.
2. Pulverize the soil using a mortar and rubber-tipped pestle, as described in Chapter 4.
3. Run a sieve analysis on the soil (Step 2), as described in Chapter 4.
4. Collect in the bottom pan the soil passing through the No. 200 sieve.
5. Wash the soil retained on the No. 200 sieve, as described in Chapter 4. Collect all the wash water and dry it in an oven.
6. Mix together the minus No. 200 portion from Step 4 and the dried minus No. 200 portion from Step 5.
7. Calculate the percent finer for the soil retained on the No. 200 sieve and above (as shown in Table 4.2).
8. Take 50 g of the minus No. 200 soil (Step 6) and run a hydrometer analysis. (Follow Steps 1 through 14 as described in Section 5.3.)
9. Report the calculations for the hydrometer analysis similar to those shown in Table 5.3. Note, however, that the percent finer now calculated (as in column 4 of Table 5.3) is

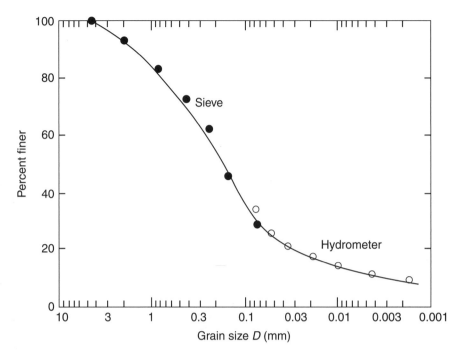

Figure 5.5. A grain-size distribution plot; combined results from sieve analysis and hydrometer analysis.

not the percent finer based on the total sample. Calculate the percent finer based on the total sample as

$$P_T = (\text{column 4 of Table 5.3}) \left(\frac{\text{percent passing No. 200 sieve}}{100} \right)$$

The percent finer passing the No. 200 sieve can be obtained from Step 7.

10. Plot a combined graph for percent finer versus grain-size distribution obtained from *both the sieve analysis and the hydrometer analysis.* An example of this is shown in Fig. 5.5. From this plot, note that there is an overlapping zone. The percent finer calculated from the sieve analysis for a given grain size does not match that calculated from the hydrometer analysis. The grain sizes obtained from a sieve analysis are the least sizes of soil grains, and the grain sizes obtained from the hydrometer analysis are the diameters of equivalent spheres of soil grains.

5.7 General Comments

A hydrometer analysis gives results from which the percent of soil finer than 0.002 mm in diameter can be estimated. It is generally accepted that the percent finer than 0.002 mm in size is clay or clay-size fractions. Most clay particles are smaller than 0.001 mm, and 0.002 mm is the upper limit. The presence of clay in a soil contributes to its plasticity.

6
Liquid Limit Test

6.1 Introduction

When a *cohesive soil* is mixed with an excessive amount of water, it will be in a somewhat *liquid state* and flow like a viscous liquid. However, when this viscous liquid is dried gradually, with the loss of moisture it will pass into a *plastic state*. With further reduction of moisture, the soil will pass into a semisolid and then into a solid state. This is shown in Fig. 6.1. The moisture content (in percent) at which the cohesive soil will pass from a liquid state to a plastic state is called the *liquid limit* of the soil. Similarly, the moisture contents (in percent) at which the soil changes from a plastic to a semisolid state and from a semisolid to a solid state are referred to as *plastic limit* and *shrinkage limit*, respectively. These limits are referred to as the *Atterberg limits* (Atterberg, 1911). In this chapter the procedure to determine the liquid limit of a cohesive soil will be discussed.

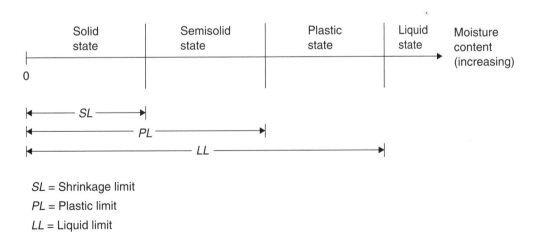

SL = Shrinkage limit
PL = Plastic limit
LL = Liquid limit

Figure 6.1. Atterberg limits.

6.2 Equipment

1. Casagrande liquid limit device
2. Grooving tool
3. Moisture cans
4. Porcelain evaporating dish
5. Spatula
6. Oven
7. Balance sensitive to 0.01 g
8. Plastic squeeze bottle
9. Paper towels

The equipment (except for the oven) is shown in Fig. 6.2.

The Casagrande liquid limit device essentially consists of a brass cup that can be raised and dropped through a distance of 10 mm (0.394 in.) on a hard rubber base by a cam operated by a crank [see Fig. 6.3(a)]. Fig. 6.3(b) shows the schematic diagram of a grooving tool.

Figure 6.2. Equipment for liquid limit test.

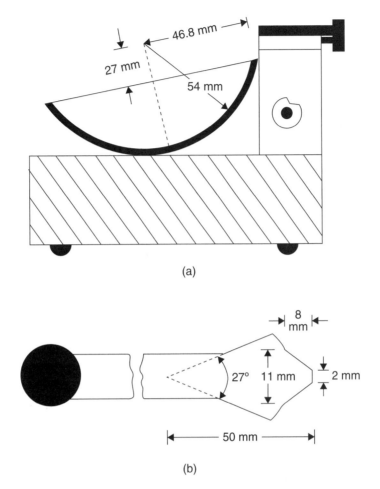

Figure 6.3. Schematic diagrams: (a) Liquid limit device. (b) Grooving tool.

6.3 Procedure

1. Determine the mass M_1 of three moisture cans.
2. Put about 250 g of air-dry soil, passed through a No. 40 sieve, into an evaporating dish. Add water from the plastic squeeze bottle and mix the soil to the form of a uniform paste.
3. Place a portion of the paste in the brass cup of the liquid limit device. Using the spatula, smooth the surface of the soil in the cup such that the maximum depth of the soil is about 8 mm.
4. Using the grooving tool, cut a groove along the centerline of the soil pat in the cup [Fig. 6.4(a).].
5. Turn the crank of the liquid limit device at the rate of about 2 revolutions per second. By this, the liquid limit cup will rise and drop through a vertical distance of 10 mm once for each revolution. The soil from the two sides of the cup will begin to flow toward

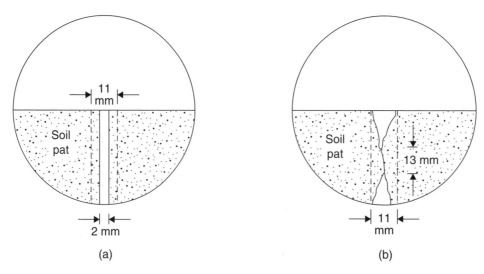

Figure 6.4. Schematic diagram (plan) of soil pat in cup of liquid limit device. (a) Beginning of test. (b) End of test.

the center. Count the number of blows N for the groove in the soil to close through a distance of 1/2 in. (13 mm), as shown in Fig. 6.4(b).

If N equals about 25 to 35, collect a moisture sample from the soil in the cup in a moisture can. Close the cover of the can and determine the mass of the can plus the moist soil, M_2.

Remove the rest of the soil paste from the cup to the evaporating dish. Use paper towels to clean the cup thoroughly.

If the soil is too dry, N will be more than about 35. In that case transfer the soil to the evaporating dish using the spatula. Clean the liquid limit cup thoroughly with paper towels. Mix the soil in the evaporating dish with more water, and try again.

If the soil is too wet, N will be less than about 25. In that case transfer the soil in the cup to the evaporating dish. Clean the liquid limit cup carefully with paper towels. Stir the soil paste with the spatula for some time to dry it up. The evaporating dish may be placed in the oven for a few minutes for drying also. *Do not* add dry soil to the wet-soil paste to reduce the moisture content in order to bring it to the proper consistency. Now try again with the liquid limit device to get a groove closure of 1/2 in. (13 mm) between 25 and 35 blows.

6. Add more water to the soil paste in the evaporating dish and mix thoroughly. Repeat Steps 3, 4, and 5 to get a groove closure of 1/2 in. (13 mm) in the liquid limit device at a blow count $N = 20$ to 25. Take a moisture sample from the cup. Transfer the rest of the soil paste to the evaporating dish. Clean the cup with paper towels.

7. Add more water to the soil paste in the evaporating dish and mix well. Repeat Steps 3, 4, and 5 to get a blow count N of between 15 and 20 for a groove closure of 1/2 in. (13 mm) in the liquid limit device. Take a moisture sample from the cup.

8. Put the three moisture cans in the oven to dry to constant masses, M_3. (The caps of the moisture cans should be removed from the top and placed at the bottom of the respective cans in the oven.)

6.4 Calculations

Determine the moisture content for each of the three trials (Steps 5, 6, and 7),

$$w(\%) = \frac{M_2 - M_3}{M_3 - M_1} \times 100 \tag{6.1}$$

6.5 Graphs

Plot a semilog graph for moisture content (arithmetic scale) versus number of blows N (log scale). This will approximate a straight line, which is called the *flow curve*. From the straight line, determine the moisture content w (%) corresponding to 25 blows. This is the *liquid limit* of the soil.

The magnitude of the slope of the flow line is called the *flow index* F_I,

$$F_1 = \frac{w_1(\%) - w_2(\%)}{\log N_2 - \log N_1} \tag{6.2}$$

Typical examples of liquid limit calculations and the corresponding graphs are shown in Table 6.1 and Fig. 6.5.

6.6 General Comments

Based on the liquid limit tests on several soils, the U.S. Army Corps of Engineers (1949) observed that the liquid limit LL of a soil can be approximately given by

$$LL = w_N \ (\%) \left(\frac{N}{25}\right)^{0.121} \tag{6.3}$$

where w_N is the moisture content, in percent, for 1/2-in. (13-mm) groove closure in the liquid limit device at N number of blows.

ASTM also recommends this equation for determining the liquid limit of soils (ASTM Test Designation D-4318). However, the value of w_N should correspond to an N value of between 20 and 30. Table 6.2 lists the values of $(N/25)^{0.121}$ for various values of N.

The presence of clay contributes to the plasticity of soil. The liquid limit of a soil will change depending on the amount and type of clay minerals present in it. The approximate ranges for the liquid limits of some clay minerals are given in Table 6.3.

Table 6.1. Liquid Limit Test

Description of soil _____ *Gray silty clay* _____ Sample no. _____ *4* _____

Location _____

Tested by _____ Date _____

Test No.	1	2	3
Can no.	8	21	25
Mass of can, M_1 (g)	15.26	17.01	15.17
Mass of can + moist soil, M_2 (g)	29.30	31.58	31.45
Mass of can + dry soil, M_3 (g)	25.84	27.72	26.96
Moisture content, $w\,(\%) = \dfrac{M_2 - M_3}{M_3 - M_1} \times 100$	32.7	36.04	38.1
Number of blows, N	35	23	17

Liquid limit LL _____ 35.2 _____

Flow index F_I _____ $\dfrac{37 - 33.7}{\log\ 30 - \log\ 20} = 18.74$ _____

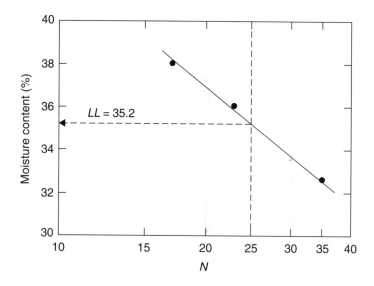

Figure 6.5. Plot of moisture content vs. number of blows for liquid limit test results reported in Table 6.1.

Table 6.2. Variation of $(N/25)^{0.121}$ Values

N	$\left(\dfrac{N}{25}\right)^{0.121}$	N	$\left(\dfrac{N}{25}\right)^{0.121}$
20	0.973	26	1.005
21	0.979	27	1.009
22	0.985	28	1.014
23	0.990	29	1.018
24	0.995	30	1.022
25	1.000		

Table 6.3. Liquid Limits of some Clay Materials

Clay Mineral	LL
Kaolinite	35–100
Illite	55–120
Montmorillonite	100–800
Chlorite	45–50
Hydrated halloysite	50–70

Casagrande (1932) concluded that each blow in a standard liquid limit device corresponds to a soil shear strength of about 0.1 kN/m^2. Hence the liquid limit of a finegrained soil gives the moisture content at which the shear strength of the soil is approximately 2.5 kN/m^2 ($\approx$ 52 lb/ft^2).

7
Plastic Limit Test

7.1 Introduction

The fundamental concept of *plastic limit* was introduced in Section 6.1 (see Fig. 6.1). Plastic limit is defined as the moisture content, in percent, at which a cohesive soil will change from a *plastic state* to a *semisolid state*. In the laboratory the *plastic limit* is defined as the moisture content, in percent, at which a thread of soil will just crumble when rolled to a diameter of 1/8 in. (3.2 mm). This test might be seen as somewhat arbitrary and, to some extent, the result may depend on the person performing the test. With practice, however, fairly consistent results may be obtained.

7.2 Equipment

1. Porcelain evaporating dish
2. Spatula
3. Plastic squeeze bottle with water
4. Moisture can
5. Ground glass plate
6. Balance sensitive to 0.01 g
7. Oven

The equipment (except for the oven) is shown in Fig. 7.1.

7.3 Procedure

1. Put approximately 20 g of a representative air-dry soil sample, passed through a No. 40 sieve, into a porcelain evaporating dish.
2. Add water from the plastic squeeze bottle to the soil and mix thoroughly.
3. Determine the mass M_1 of a moisture can in grams and record it on the data sheet.

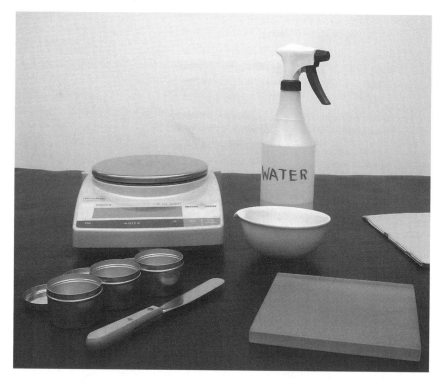

Figure 7.1. Equipment for plastic limit test.

4. From the moist soil prepared in Step 2, prepare several ellipsoidally shaped soil masses by squeezing the soil with your fingers.

5. Take one of the ellipsoidally shaped soil masses (Step 4) and roll it on a ground glass plate using the palm of your hand (Fig. 7.2). The rolling should be done at the rate of about 80 strokes per minute. Note that one complete backward and one complete forward motion of the palm constitute a stroke.

6. When the thread being rolled in Step 5 reaches 1/8 in. (3.2 mm) in diameter, break it up into several small pieces and squeeze one piece with your fingers to form again an ellipsoidal mass.

7. Repeat Steps 5 and 6 until the thread crumbles into several pieces when it reaches a diameter of 1/8 in. (3.2 mm). It is possible that a thread may crumble at a diameter larger than 1/8 in. (3.2 mm) during a given rolling process, whereas it did not crumble at the same diameter during the immediately previous rolling.

8. Collect the small crumbled pieces in the moisture can and put the cover on the can.

9. Take the other ellipsoidal soil masses formed in Step 4 and repeat Steps 5 through 8.

10. Determine the mass of the moisture can plus the wet soil, M_2, in grams. Remove the cap from the top of the can and place the can in the oven (with the cap at the bottom of the can).

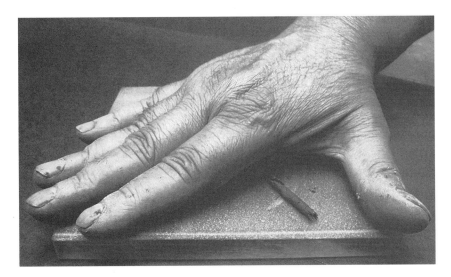

Figure 7.2. Ellipsoidal soil mass being rolled in a thread on a glass plate.

11. After about 24 hours, remove the can from the oven and determine the mass of the can plus the dry soil, M_3, in grams.

7.4 Calculations

Calculate the plastic limit *PL*,

$$PL = \frac{\text{mass of moisture}}{\text{mass of dry soil}} = \frac{M_2 - M_3}{M_3 - M_1} \times 100 \tag{7.1}$$

The results may be presented in tabular form, as shown in Table 7.1. If the liquid limit of the soil is known, calculate the *plasticity index PI*

$$PI = LL - PL \tag{7.2}$$

7.5 General Comments

Table 7.2 lists the approximate ranges for the plastic limits of some clay minerals. The liquid limit and the plasticity index of cohesive soils are important parameters for classification purposes. The engineering soil classification systems are described in Chapter 9. The plasticity index is also used to determine the activity *A* of a clayey soil, which is defined as

$$A = \frac{PI}{\% \text{ clay-size fraction (by weight)}}$$

The range of activity *A* of some clay minerals is given in Table 7.3.

Table 7.1. Plastic Limit Test

Description of soil _____ *Gray clayey silt* _____ Sample no. ___3___

Location _____

Tested by _____ Date _____

Can No.	103
Mass of can, M_1 (g)	13.33
Mass of can + moist soil, M_2 (g)	23.86
Mass of can + dry soil, M_3 (g)	22.27
$PL = \dfrac{M_2 - M_3}{M_3 - M_1} \times 100$	17.78

Plasticity index $PI = LL - PL = $ ___34 − 17.78___ = ___18.78___

Table 7.2. Range of Plastic Limits

Clay Mineral	*PL*
Kaolinite	20–40
Illite	35–60
Montmorillonite	50–100
Chlorite	35–40
Hydrated halloysite	40–60

Table 7.3. Range of *A* Values

Clay Mineral	*A*
Kaolinite	0.3–0.5
Illite	0.5–1.2
Montmorillonite	1.5–7.0
Hydrated halloysite	0.1–0.2

Liquidity index LI is a term that defines the relative consistency of a clayey soil in its natural state. It can be expressed as

$$LI = \frac{w - PL}{LL - PL}$$

where w is the natural moisture content.

For sensitive clays LI may be greater than 1. Heavily overconsolidated clays may have $LI < 0$.

8
Shrinkage Limit Test

8.1 Introduction

The fundamental concept of shrinkage limit was presented in Fig. 6.1. A saturated clayey soil, when gradually dried, will lose moisture, and subsequently there will be a reduction in the volume of the soil mass. During the drying process a condition will be reached when any further drying will result in a reduction in moisture content without any decrease in volume (Fig. 8.1). The moisture content of the soil, in percent, at which the decrease in soil volume ceases is defined as the *shrinkage limit*.

8.2 Equipment

1. Shrinkage limit dish, usually made of porcelain, about 1.75 in. (44.4 mm) in diameter and 0.5 in. (12.7 mm) high
2. Glass cup, 2.25 in. (57.2 mm) in diameter and 1.25 in. (31.8 mm) high
3. Glass plate with three prongs
4. Porcelain evaporating dish about 5.5 in. (139.7 mm) in diameter
5. Spatula
6. Plastic squeeze bottle with water
7. Steel straightedge
8. Mercury
9. Watch glass
10. Balance sensitive to 0.01 g

This equipment is shown in Fig. 8.2. A schematic diagram of the glass plate with three metal prongs is shown in Fig. 8.3.

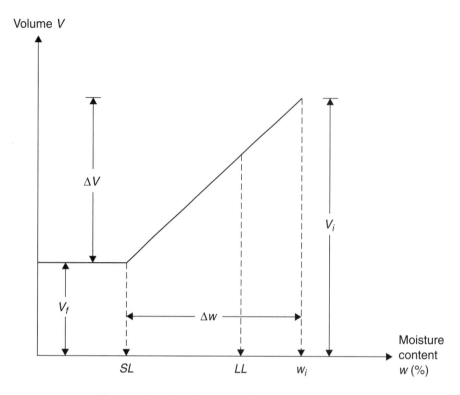

Figure 8.1. Definition of shrinkage limit.

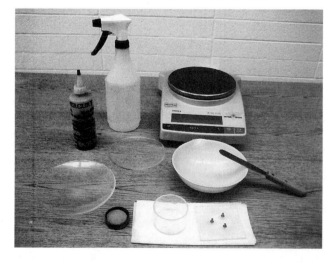

Figure 8.2. Equipment needed for determination of shrinkage limit.

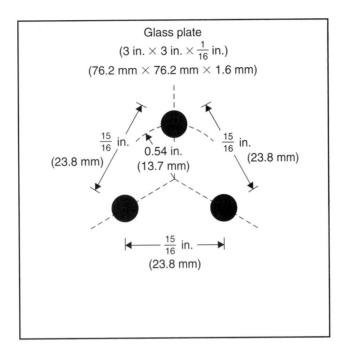

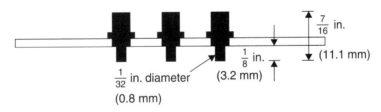

Figure 8.3. Glass plate with three metal prongs.

8.3 Procedure

1. Put about 80–100 g of a representative air-dry soil, passed through a No. 40 sieve, into an evaporating dish.
2. Add water to the soil from the plastic squeeze bottle and mix it thoroughly into the form of a creamy paste. Note that the moisture content of the paste should be above the liquid limit of the soil to ensure full saturation.
3. Coat the shrinkage limit dish lightly with petroleum jelly and then determine the mass M_1, in grams, of the coated dish.
4. Fill the dish about one-third full with the soil paste. Tap the dish on a firm surface so that the soil flows to the edges of the dish and no air bubbles exist.
5. Repeat Step 4 until the dish is full.
6. Level the surface of the soil with the steel straightedge. Clean the sides and bottom of the dish with paper towels.
7. Determine the mass of the dish plus the wet soil, M_2, in grams.

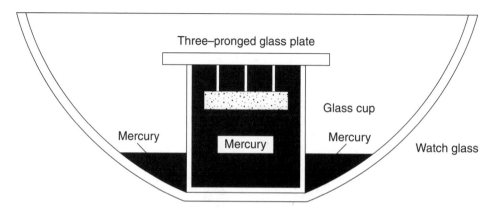

Figure 8.4. Determination of the volume of the soil pat (Step 12).

8. Allow the dish to air dry (about 6 hours) until the color of the soil pat becomes lighter. Then put the dish with the soil into the oven to dry.
9. Determine the mass of the dish and the oven-dry soil pat, M_3, in grams.
10. Remove the soil pat from the dish.
11. In order to find the volume V_i of the shrinkage limit dish, fill the dish with mercury. (*Note:* The dish should be placed on a watch glass.) Use the three-pronged glass plate and level the surface of the mercury in the dish. The excess mercury will flow into the watch glass. Determine the mass of mercury in the dish, M_4, in grams.
12. In order to determine the volume V_f of the dry soil pat, fill the glass cup with mercury. (The cup should be placed on a watch glass.) Using the three-pronged glass plate, level the surface of the mercury in the glass cup. Remove the excess mercury on the watch glass. Place the dry soil pat on the mercury in the glass cup. The soil pat will float. Now, using the three-pronged glass plate, slowly push the soil pat into the mercury until the soil pat is completely submerged (Fig. 8.4). The displaced mercury will flow out of the glass cup and will be collected on the watch glass. Determine the mass of the displaced mercury on the watch glass, M_5, in grams.

8.4 Calculations

1. Calculate the initial moisture content of the soil at molding,

$$w_i \, (\%) = \frac{M_2 - M_3}{M_3 - M_1} \times 100 \tag{8.1}$$

2. Calculate the change in moisture content (%) before the volume reduction ceased (refer to Fig. 8.1).

$$\Delta w_i \, (\%) = \frac{\left(V_i - V_f\right) \rho_w}{\text{mass of dry soil pat}} = \frac{M_4 - M_5}{13.6(M_3 - M_1)} \times 100 \tag{8.2}$$

where ρ_w is the density of water, $= 1 \, \text{g/cm}^3$.

Table 8.1. Shrinkage Limit Test

Description of soil _____ *Dark brown clayey silt* _____ Sample no. _____ *8* _____

Location _____ *Westwind Boulevard* _____

Tested by _____ Date _____

Test No.	*1*	
Mass of coated shrinkage limit dish, M_1 (g)	*12.34*	
Mass of dish + wet soil, M_2 (g)	*40.43*	
Mass of dish + dry soil, M_3 (g)	*33.68*	
$w_i\,(\%) = \dfrac{M_2 - M_3}{M_3 - M_1} \times 100$	*31.63*	
Mass of mercury to fill dish, M_4 (g)	*198.83*	
Mass of mercury displaced by soil pat, M_5 (g)	*150.30*	
$\Delta w_i\,(\%) = \dfrac{M_4 - M_5}{13.6\,(M_3 - M_1)} \times 100$	*16.72*	
$SL = w_i - \dfrac{M_4 - M_5}{13.6\,(M_3 - M_1)} \times 100$	*14.91*	

3. Calculate the shrinkage limit,

$$SL = w_i - \frac{M_4 - M_5}{13.6\,(M_3 - M_1)} \times 100 \tag{8.3}$$

Note that M_4 and M_5 are in grams, and the specific gravity of the mercury is 13.6. A sample calculation is shown in Table 8.1.

8.5 General Comments

The ratio of the liquid limit to the shrinkage limit (LL/SL) of a soil gives a good idea about its shrinkage properties. If the ratio LL/SL is large, the soil in the field may undergo an undesirable volume change due to a change in moisture. New foundations constructed on these soils may show cracks due to shrinking and swelling of the soil that result from seasonal moisture change.

Another parameter, called *shrinkage ratio* (*SR*), may also be determined from the shrinkage limit test. Referring to Fig. 8.1,

$$SR = \frac{\Delta V / V_f}{\Delta w / M_s} = \frac{\Delta V / V_f}{\Delta V \, \rho_w / M_s} = \frac{M_s}{\rho_w V_f} \tag{8.4}$$

where M_s is the mass of the dry soil pat,

$$M_s = M_3 - M_1$$

If M_s is in grams, V_f is in cm^3, and $\rho_w = 1$ g/cm^3. Thus,

$$SR = \frac{M_s}{V_f} \tag{8.5}$$

The shrinkage ratio gives an indication of the volume change with a change in moisture content.

9
Engineering Classification of Soils

9.1 Introduction

Soils vary widely in their grain-size distributions (Chapters 4 and 5). Also, depending on the type and quantity of clay minerals present, the plastic properties of soils (Chapters 6, 7, and 8) may be very different. Various types of engineering works require the identification and classification of soil in the field. In the design of foundations and earth-retaining structures, construction of highways, and so on, it is necessary for soils to be arranged in specific groups and/or subgroups based on their grain-size distribution and plasticity. The process of placing soils into various groups and/or subgroups is called *soil classification.*

For engineering purposes there are two major systems presently used in the United States. They are (1) the *American Association of State Highway and Transportation Officials* (AASHTO) *Classification System* and (2) the *Unified Classification System.* These two systems will be discussed in this chapter.

9.2 American Association of State Highway and Transportation Officials (AASHTO) System of Classification

The AASHTO classification system was originally initiated by the Highway Research Board (now called Transportation Research Board) in 1943. This classification system, which has undergone several changes since then, is presently used by federal, state, and county highway departments in the United States. In this soil classification system, soils are generally placed in seven major groups: A-1, A-2, A-3, A-4, A-5, A-6, and A-7. Group A-1 is divided into two subgroups (A-1-a and A-1-b). Group A-2 is divided into four subgroups (A-2-4, A-2-5,

A-2-6, and A-2-7). Soils in group A-7 are also divided into two subgroups (A-7-5 and A-7-6). This system is also included in ASTM test designation D-3282.

Along with these soil groups and subgroups another factor, called the *group index* (*GI*), is also included in this system. The importance of the group index can be explained as follows. Let us assume that two soils fall under the same group; however, they may have different values of *GI*. The soil that has a lower value of group index is likely to perform better as a highway subgrade material.

The procedure for classifying soil under the AASHTO system is outlined in the following.

9.3 Step-by-Step Procedure for AASHTO Classification

1. Determine the percentage of soil passing through No. 200 U.S. sieve (0.075-mm opening).
   a. If 35% or less passes the No. 200 sieve, it is a coarse-grained material. Proceed to Steps 2 and 4.
   b. If more than 35% passes the No. 200 sieve, it is a fine-grained material (i.e., silty or clayey). For this, go to Steps 3 and 5.

Determination of Groups or Subgroups

2. For coarse-grained soils, determine the percent passing U.S. sieve Nos. 10, 40, and 200 and, in addition, the liquid limit and plasticity index. Then proceed to Table 9.1. Start from the top line and compare the known soil properties with those given in the table (columns 2 through 6). Go down one line at a time until a line is found for which all the properties of the desired soil match. The soil group (or subgroup) is determined from column 1.
3. For fine-grained soils, determine the liquid limit and the plasticity index. Then go to Table 9.2. Start from the top line. By matching the soil properties from columns 2, 3, and 4, determine the proper soil group (or subgroup).

Determination of Group Index

4. To determine the group index *GI* of coarse-grained soils, the following rules need to be observed.
   a. *GI* for soils in groups (or subgroups) A-1-a, A-1-b, A-2-4, A-2-5, and A-3 is zero.
   b. For *GI* in soils in groups A-2-6 and A-2-7 use the following equation:

$$GI = 0.01 \, (F_{200} - 15) \, (PI - 10) \tag{9.1}$$

   where F_{200} = percent passing No. 200 sieve
   PI = plasticity index

   If *GI* comes out negative, round it off to zero. If *GI* is positive, round it off to the nearest whole number.

Table 9.1. AASHTO Classification for Coarse-Grained Soils

| Soil Group (1) | | Grain Size | | | Liquid Limit* (5) | Plasticity Index* (6) | Material Type (7) | Subgrade Rating (8) |
		Passing No. 10 Sieve (2)	Passing No. 40 Sieve (3)	Passing No. 200 Sieve (4)				
A-1	A-1-a	50 max.	30 max.	15 max.		6 max.	Stone fragments, gravel, and sand	Excellent to good
	A-1-b		50 max.	25 max.		6 max.		
A-3			51 min.	10 max.		Nonplastic	Fine sand	
A-2	A-2-4			35 max.	40 max.	10 max.	Silty and clayey gravel and sand	
	A-2-5			35 max.	41 min.	10 max.		
	A-2-6			35 max.	40 max.	11 min.		
	A-2-7			35 max.	41 min.	11 min.		

* Based on fraction passing No. 40 U.S. sieve.

59

Table 9.2. AASHTO Classification for Fine-Grained Soils

Soil Group (1)		Passing No. 200 Sieve (2)	Liquid Limit* (3)	Plasticity Index* (4)	Material Type (5)	Subgrade Rating (6)
A-4		36 min.	40 max.	10 max.	Silty soil	Fair to poor
A-5		36 min.	41 min.	10 max.	Silty soil	Fair to poor
A-6		36 min.	40 max.	11 min.	Clayey soil	Fair to poor
A-7	A-7-5	36 min.	41 min.	11 min. and $PI \leq LL - 30$	Clayey soil	Fair to poor
	A-7-6	36 min.	41 min.	11 min. and $PI > LL - 30$	Clayey soil	Fair to poor

* Based on fraction passing No. 40 U.S. sieve.

60

5. To obtain *GI* for fine-grained soils, use the following equation:

$$GI = (F_{200} - 35)\,[0.2 + 0.005\,(LL - 40)] + 0.01\,(F_{200} - 15)\,(PI - 10) \qquad (9.2)$$

If *GI* comes out negative, round it off to zero; if it is positive, round it off to the nearest whole number.

Soil Classification

6. The final classification of a soil is given by first writing down the group, or subgroup followed by the group index in parentheses.

9.4 Examples

Figure 9.1 shows the range of *PI* and *LL* for soil groups A-2-4, A-2-5, A-2-6, A-2-7, A-4, A-5, A-6, A-7-5, and A-7-6.

Example 9.1

The following are the characteristics of two soils. Classify the soils according to the AASHTO system.

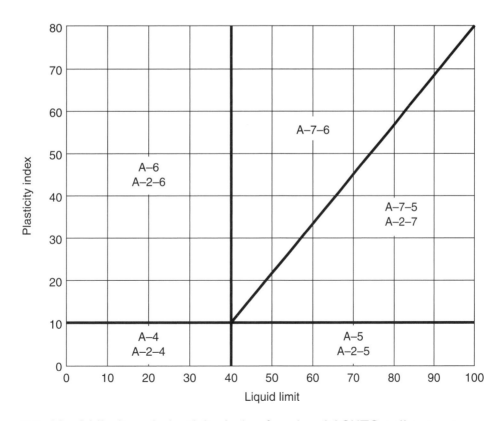

Figure 9.1. Liquid limit and plasticity index for nine AASHTO soil groups.

Soil A:
Percent passing No. 4 sieve   = 98
Percent passing No. 10 sieve   = 90
Percent passing No. 40 sieve   = 76
Percent passing No. 200 sieve = 34
Liquid limit   = 38
Plastic limit   = 26

Soil B:
Percent passing No. 4 sieve   = 100
Percent passing No. 10 sieve   = 98
Percent passing No. 40 sieve   = 86
Percent passing No. 200 sieve = 58
Liquid limit   = 49
Plastic limit   = 28

Solution

Refer to Section 9.3.

Soil A:

1. The soil has 34% (which is less than 35%) passing through No. 200 sieve. So this is a coarse-grained soil.
2. For this soil the liquid limit is 38. From Eq. (7.2),

$$PI = LL - PL = 38 - 26 = 12$$

From Table 9.1, by matching, the soil is found to belong to subgroup A-2-6.

3. From Eq. (9.1),

$$GI = 0.01 \, (F_{200} - 15) \, (PI - 10)$$
$$= 0.01 \, (34 - 15) \, (12 - 10) = 0.01 \times 19 \times 2$$
$$= 0.38 \approx 0$$

So the soil can be classified as A-2-6(0).

Soil B:

1. The soil has 58% (which is more than 35%) passing through No.200 sieve. So this is a fine-grained soil.
2. The liquid limit of the soil is 49. From Eq. (7.2),

$$PI = LL - PL = 49 - 28 = 21$$

3. From Table 9.2 the soil is either A-7-5 or A-7-6. However, for this soil,

$$PI = 21 > LL - 30 = 49 - 30 = 19$$

So this soil is A-7-6.

4. From Eq. (9.2),

$$GI = (F_{200} - 35) [0.2 + 0.005 (LL - 40)] + 0.01 (F_{200} - 15) (PI - 10)$$
$$= (58 - 35) [0.2 + 0.005 (49 - 40)] + 0.01 (58 - 15) (21 - 10)$$
$$= 5.64 + 4.73 = 10.37 \approx 10$$

So the soil can be classified as A-7-6(10).

9.5 Unified Classification System

This classification system was originally developed in 1942 by Arthur Casagrande for airfield construction during World War II. The work was conducted on behalf of the U.S. Army Corps of Engineers. At a later date, with the cooperation of the United States Bureau of Reclamation, the classification was modified. More recently the American Society for Testing and Materials (ASTM) introduced a more definite system for the group names of soils. In the present form it is widely used by foundation engineers all over the world. Unlike the AASHTO system, the unified system uses symbols to represent the soil types and the index properties of a soil. They are given in Table 9.3.

Soil groups are developed by combining the symbols for the two categories listed in Table 9.3, such as GW, SM, and so forth.

Table 9.3. Unified Soil Classification System Symbols

Symbol	Soil Type	Symbol	Index Property
G	Gravel	W	Well graded (for grain-size distribution)
S	Sand	P	Poorly graded (for grain-size distribution)
M	Silt	L	Low to medium plasticity
C	Clay	H	High plasticity
O	Organic silts and clays		
Pt	Highly organic soil and peat		

9.6 Step-by-Step Procedure for Unified Classification System

1. If it is peat (i.e., primarily organic matter, dark in color, having organic odor), classify it as Pt by visual observation. For all other soils, determine the percent of soil passing through a No. 200 U.S. sieve (F_{200}).

2. Determine the percent retained on the No. 200 U.S. sieve (R_{200}) as

$$R_{200} = 100 - F_{200}$$
$$\uparrow$$
$$\text{(nearest whole number)}$$

3. If R_{200} is greater than 50%, it is a coarse-grained soil. However, if R_{200} is less than or equal to 50%, it is a fine-grained soil. For the case where $R_{200} \leq 50\%$ (fine-grained soil), go to Step 4. If $R_{200} > 50\%$, go to Step 5.

4. For fine-grained soils ($R_{200} \leq 50\%$), determine whether the soil is organic or inorganic in nature. If the soil is organic, the group symbol can be OH or OL. If the soil is inorganic, the group symbol can be CL, ML, CH, MH, or CL-ML. To determine the group symbols, go to Table 9.4 and Fig. 9.2. Table 9.5 gives a general description of the type of soil within each group.

5. For coarse-grained soils, determine the percent retained on the No. 4 U.S. sieve (R_4),

$$R_4 = 100 - F_4$$
$$\uparrow$$
$$\text{(nearest whole number)}$$

where F_4 is percent finer than the No. 4 sieve. Note that R_4 is the percent gravel fraction *GF* in the soil,

$$GF = R_4 \tag{9.3}$$

a. If $R_4 > 0.5R_{200}$ it is a gravelly soil. These soils may have the following group symbols:

GW   GW–GM
GP   GW–GC
GM   GP–GM
GC   GP–GC
       GC–GM

Determine the following:
(1) F_{200}
(2)   Uniformity coefficient, $C_u = D_{60}/D_{10}$ (see Chapter 4)

Table 9.4. Unified Soil Classification
System—Group Symbols for Silty and Clayey
Fine-Grained Soils ($R_{200} \leq 50$)

Group Symbol	Criteria
CL	Inorganic; $LL < 50$; $PI > 7$; Atterberg limits plot on or above A line (see CL zone in Fig. 9.2)
ML	Inorganic; $LL < 50$; $PI < 4$; Atterberg limits plot below A line (see ML zone in Fig. 9.2)
OL	Organic; $LL_{\text{oven dried}}/LL_{\text{not dried}} < 0.75$; $LL < 50$ (see OL zone in Fig. 9.2)
CH	Inorganic; $LL \geq 50$; Atterberg limits plot on or above A line (see CH zone in Fig. 9.2)
MH	Inorganic; $LL \geq 50$; Atterberg limits plot below A line (see MH zone in Fig. 9.2)
OH	Organic; $LL_{\text{oven dried}}/LL_{\text{not dried}} < 0.75$; $LL \geq 50$ (see OH zone in Fig. 9.2)
CL–ML	Inorganic; Atterberg limits plot in hatched zone in Fig. 9.2
Pt	Peat, muck, and other highly organic soils

(3) Coefficient of gradation, $C_c = (D_{30})^2 / (D_{10} \times D_{60})$

(4) LL (of minus No. 40 sieve)

(5) PI (of minus No. 40 sieve)

Go to Table 9.6 to find the group symbols. Table 9.7 gives a general description of the type of soil within each group.

b. If $R_4 \leq 0.5\, R_{200}$, it is a sandy soil. These soils may have the following group symbols:

SW   SW–SM
SP   SW–SC
SM   SP–SM
SC   SP–SC
       SM–SC

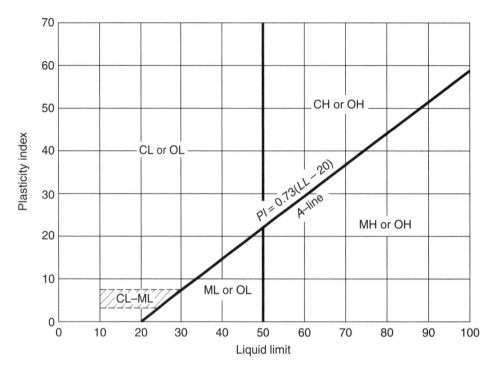

Figure 9.2. Plasticity chart for group symbols of fine-grained soils.

Table 9.5. General Description of Soil Types for Group Symbols in Table 9.4

Group Symbol	General Description
CL	Inorganic clays of low to medium plasticity, gravelly clays, sandy clays, silty clays, lean clays
ML	Inorganic silts, very fine sands, rock flour, silty or clayey fine sands
OL	Organic silts and organic silty clays of low plasticity
CH	Inorganic clays of high plasticity, fat clays
MH	Inorganic silts, micaceous or diatomaceous fine sands or silts, elastic silts
OH	Organic clays of medium to high plasticity

Table 9.6. Unified Soil Classification System—Group Symbols for Gravelly Soils ($R_{200} > 50\%$ and $R_4/R_{200} > 0.5$)

Group Symbol	Criteria
GW	Less than 5% passing No. 200 sieve; $C_u = D_{60}/D_{10} \geq 4$; $C_c = (D_{30})^2/(D_{10} \times D_{60})$ between 1 and 3
GP	Less than 5% passing No. 200 sieve; not meeting both criteria for GW
GM	More than 12% passing No. 200 sieve; Atterberg limits plot below *A* line (Fig. 9.2) or *PI* < 4
GC	More than 12% passing No. 200 sieve; Atterberg limits plot above *A* line (Fig. 9.2) or *PI* > 7
GC–GM	More than 12% passing No. 200 sieve; Atterberg limits fall in hatched area marked CL–ML in Fig. 9.2
GW–GM	5–12% passing No. 200; meets criteria for GW and GM
GW–GC	5–12% passing No. 200 sieve; meets criteria for GW and GC
GP–GM	5–12% passing No. 200 sieve; meets criteria for GP and GM
GP–GC	5–12% passing No. 200 sieve; meets criteria for GW and GC

Table 9.7. General Description of Soil Types for Group Symbols in Table 9.6

Group Symbol	General Description
GW	Well-graded gravels and gravel–sand mixtures, little or no fines
GP	Poorly graded gravels and gravel–sand mixtures, little or no fines
GM	Silty gravels, gravel–sand–silt mixtures
GC	Clayey gravels, gravel–sand–clay mixtures

Determine the following:

(1) F_{200}
(2) Uniformity coefficient, $C_u = D_{60}/D_{10}$ (see Chapter 4)
(3) Coefficient of gradation, $C_c = (D_{30})^2/(D_{10} \times D_{60})$
(4) *LL* (of minus No. 40 sieve)
(5) *PI* (of minus No. 40 sieve)

Go to Table 9.8 to find the group symbols. Table 9.9 gives a general description of the soil type within each group.

Table 9.8.   Unified Soil Classification System—Group Symbols for Sandy Soils ($R_{200} > 50\%$ and $R_4/R_{200} \leq 0.5$)

Group Symbol	Criteria
SW	Less than 5% passing No. 200 sieve; $C_u = D_{60}/D_{10} \geq 6$; $C_c = (D_{30})^2/(D_{10} \times D_{60})$ between 1 and 3
SP	Less than 5% passing No. 200 sieve; not meeting both criteria for SW
SM	More than 12% passing No. 200 sieve; Atterberg limits plot below *A* line (Fig. 9.2) or *PI* < 4
SC	More than 12% passing No. 200 sieve; Atterberg limits plot above *A* line (Fig. 9.2) or *PI* > 7
SC–SM	More than 12% passing No. 200 sieve; Atterberg limits fall in hatched area marked CL–ML in Fig. 9.2
SW–SM	5–12% passing No. 200 sieve; meets criteria for SW and SM
SW–SC	5–12% passing No. 200 sieve; meets criteria for SW and SC
SP–SM	5–12% passing No. 200 sieve; meets criteria for SP and SM
SP–SC	5–12% passing No. 200 sieve; meets criteria for SW and SC

Table 9.9. General Description of Soil Types for Group Symbols in Table 9.8

Group Symbol	General Description
SW	Well-graded sands and gravelly sands, little or no fines
SP	Poorly graded sands and gravely sands, little or no fines
SM	Silty sands, sand–silt mixtures
SC	Clayey sands, sand–clay mixtures

9.7 Examples

Example 9.2

Classify Soils A and B as given in Example 9.1 and obtain the group symbols. Assume Soil B to be inorganic.

Soil A:
Percent passing No. 4 sieve   = 98
Percent passing No. 10 sieve   = 90
Percent passing No. 40 sieve   = 76
Percent passing No. 200 sieve = 34
Liquid limit   = 38
Plastic limit   = 26

Soil B:
Percent passing No. 4 sieve   = 100
Percent passing No. 10 sieve   = 98
Percent passing No. 40 sieve   = 86
Percent passing No. 200 sieve = 58
Liquid limit   = 49
Plastic limit   = 28

Solution

Refer to Section 9.6.

Soil A:

1. $F_{200} = 34\%$
2. $R_{200} = 100 - F_{200} = 100 - 34 = 66\%$
3. $R_{200} = 66\% > 50\%$. So it is a coarse-grained soil.
4. Skip Step 4.
5. $R_4 = 100 - F_4 = 2\%$

   $R_4 < 0.5R_{200} = 33\%$. So it is a sandy soil. $F_{200} > 12\%$. Thus C_u and C_c values are not needed.

$$PI = LL - PL = 38 - 26 = 12$$

Note that the actual PI is less than $0.73(LL - 20) = 0.73(38 - 20) = 13.14$.

From Table 9.8, the group symbol is SM.

Soil B:

1. $F_{200} = 58\%$
2. $R_{200} = 100 - F_{200} = 100 - 58 = 42\%$
3. $R_{200} = 42\% < 50\%$. So it is a fine-grained soil.
4. From Table 9.4, $LL = 49 < 50$.

$$PI = 49 - 28 = 21$$

$$PI = 21 < 0.73(LL - 20) = 0.73(49 - 20) = 21.17$$

So the group symbol is ML.

10
Constant-Head Permeability Test in Sand

10.1 Introduction

The rate of flow of water through a soil specimen of gross cross-sectional area A can be expressed as

$$q = kiA \qquad (10.1)$$

where $q =$ flow in unit time
$k =$ coefficient of permeability
$i =$ hydraulic gradient

Table 10.1 lists the general ranges of the coefficient of permeability k for various types of soil.

The coefficient of permeability of sand can be determined easily in the laboratory by two simple methods, (1) the constant-head test and (2) the variable-head test. In this chapter the *constant-head test method* will be discussed.

10.2 Equipment

1. Constant-head permeameter
2. Graduated cylinder (250 cm^3 or 500 cm^3)
3. Balance sensitive to 0.1 g
4. Thermometer sensitive to 0.1°C
5. Rubber tubing
6. Stopwatch

Table 10.1. Range of k for Various Soil Types

Soil	k (cm/s)
Clean gravel	$10^2 - 10^0$
Coarse sand	$10^0 - 10^{-2}$
Fine sand	$10^{-2} - 10^{-3}$
Silty clay	$10^{-3} - 10^{-5}$
Clay	Less than 10^{-6}

10.3 Constant-Head Permeameter

A schematic diagram of a constant-head permeameter is shown in Fig. 10.1. This instrument can be assembled in the laboratory at very low cost. It essentially consists of a plastic soil specimen cylinder, two porous stones, two rubber stoppers, one spring, one constant-head chamber, a large funnel, a stand, a scale, three clamps, and some plastic tubes. The plastic cylinder may have an inside diameter of 2.5 in. (63.5 mm). This is because 2.5-in. (63.5-mm)-diameter porous stones are usually available in most soil laboratories. The length of the specimen tube may be about 12 in. (304.8 mm).

It is important to keep in mind that the minimum inside diameter of the specimen cylinder should be about 8 to 12 times the maximum particle size of the soil to be tested (ASTM, 2007). Table 10.2 gives some recommended diameters of specimen cylinders.

10.4 Procedure

1. Determine the mass of the plastic specimen tube, the porous stones, the spring, and the two rubber stoppers, M_1.
2. Slip the bottom porous stone into the specimen tube and then fix the bottom rubber stopper to the specimen tube.
3. Collect oven-dry sand in a container. Using a spoon, pour the sand into the specimen tube in small layers, and compact it by vibration and/or other compacting means. (*Note:* By changing the degree of compaction, a number of test specimens having different void ratios can be prepared.)
4. When the length of the specimen tube is about two-thirds the length of the tube, slip the top porous stone into the tube to rest firmly on the specimen.
5. Place a spring on the top porous stone, if necessary.
6. Fix a rubber stopper to the top of the specimen tube. (*Note:* The spring in the assembled position will not allow any expansion of the specimen volume, and thus the void ratio, during the test).

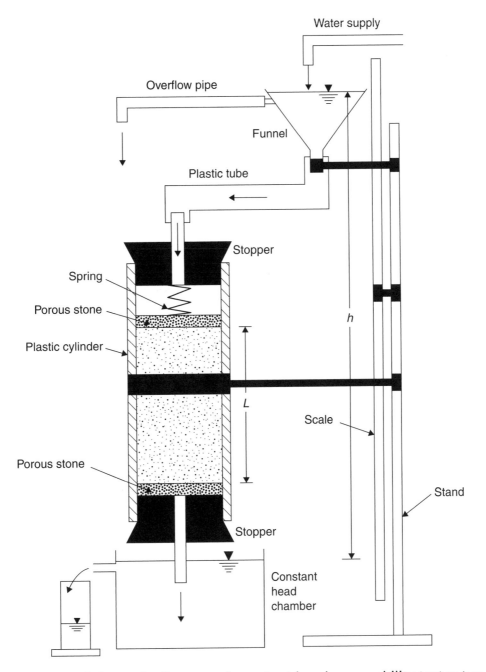

Figure 10.1. Schematic diagram of constant-head permeability test setup.

7. Determine the mass of the assembly, M_2 (Step 6).
8. Measure the length L of the compacted specimen in the tube.
9. Assemble the permeameter near a sink, as shown in Fig. 10.1.
10. Run water into the top of the large funnel fixed to the stand through a plastic tube from the water inlet. The water will flow through the specimen to the constant-head chamber.

Table 10.2.   Recommended Inside Diameters of Specimen Cylinders*

Maximum Particle Size Range	Sieve Analysis	Minimum Cylinder Diameter
No. 10 (2-mm) sieve to 3/8-in. (9.5-mm) sieve	Less than 35% of total soil retained on No. 10 sieve	3 in. (76 mm)
	More than 35% of total soil retained on No. 10 sieve	4.5 in. (114 mm)
3/8-in. (9.5-mm) sieve to 3/4-in. (19.0-mm) sieve	Less than 35% of total soil retained on 3/8-in. (9.5-mm) sieve	6 in. (152 mm)
	More than 35% of total soil retained on 3/8-in. (9.5-mm) sieve	9 in. (229 mm)

* ASTM (2007).

After some time the water will flow into the sink through the outlet in the constant-head chamber. (*Note:* Make sure that water does not leak from the specimen tube.)

11. Adjust the supply of water to the funnel so that the water level in the funnel remains constant. At the same time, allow the flow to continue for about 10 minutes in order to saturate the specimen. (*Note:* Some air bubbles may appear in the plastic tube connecting the funnel to the specimen tube. Remove the air bubbles.)

12. After a steady flow is established (that is, once the head difference h becomes constant), collect the water Q flowing out of the constant-head chamber in a graduated cylinder. Record the collection time t with a stopwatch.

13. Repeat Step 12 three times. Keep the collection time t the same and determine Q. Then find the average value of Q.

14. Change the head difference h and repeat Steps 11, 12, and 13 about three times.

15. Record the temperature T of the water to the nearest degree. (*Note:* This value is sufficiently accurate for this type of test.)

10.5 Calculations

1. Calculate the void ratio of the compacted specimen as follows. The dry density ρ_d of the soil specimen is

$$\rho_d = \frac{M_2 - M_1}{(\pi/4)D^2 L}$$

Thus,

$$e = \frac{G_s \rho_w}{\rho_d} - 1 \tag{10.2}$$

where G_s = specific gravity of soil solids
ρ_w = density of water
D = diameter of specimen
L = length of specimen

2. Calculate k,

$$k = \frac{QL}{Aht} \tag{10.3}$$

where A is the area of the specimen,

$$A = \frac{\pi}{4}D^2$$

3. The value of k is usually given for water at a test temperature of 20°C. So calculate $k_{20°C}$,

$$k_{20°C} = k_{T°C}\frac{\eta_{T°C}}{\eta_{20°C}} \tag{10.4}$$

where $\eta_{T°C}$ and $\eta_{20°C}$ are the viscosities of water at $T°C$ and 20°C, respectively. Table 10.3 gives the values of $\eta_{T°C}/\eta_{20°C}$ for various values of T (in °C).

Tables 10.4 and 10.5 show sample calculations for the permeability test.

10.6 General Comments

Several relations between k and the void ratio e for sandy soils have been proposed. They are of the form

$$k \propto e^2 \tag{10.5}$$

$$k \propto \frac{e^2}{1+e} \tag{10.6}$$

$$k \propto \frac{e^3}{1+e} \tag{10.7}$$

It is important, however, to point out that these relationships are approximate, and the actual value of k may vary widely.

Table 10.3.   Variation of $\eta_{T^\circ C}/\eta_{20^\circ C}$

Temperature $T(^\circ C)$	$\eta_{T^\circ C}/\eta_{20^\circ C}$	Temperature $T(^\circ C)$	$\eta_{T^\circ C}/\eta_{20^\circ C}$
15	1.135	23	0.931
16	1.106	24	0.910
17	1.077	25	0.889
18	1.051	26	0.869
19	1.025	27	0.850
20	1.000	28	0.832
21	0.976	29	0.814
22	0.953	30	0.797

Table 10.4.   Determination of Void Ratio of Specimen—Constant-Head Permeability Test

Description of soil _____ *Uniform sand* _____   Sample no. _____

Location _____

Length of specimen L ____ *13.2* ____ cm   Diameter of specimen D _____ *6.35* _____ cm

Tested by _____   Date _____

Volume of specimen, $V = \dfrac{\pi}{4}D^2L \, (\text{cm}^3)$	*418.03*
Specific gravity of soil solids, G_s	*2.66*
Mass of specimen tube with fittings, M_1 (g)	*238.4*
Mass of tube with fittings and specimen, M_2 (g)	*965.3*
Dry density of specimen, $\rho_d = \dfrac{M_2 - M_1}{V} \, (\text{g/cm}^3)$	*1.74*

Void ratio of specimen $e = \dfrac{G_s \rho_w}{\rho_d} - 1 = $ ___ *0.53* ___

(*Note:* $\rho_w = 1 \, \text{g/cm}^3$)

Table 10.5. Determination of Coefficient of Permeability—Constant-Head Permeability Test

Description of soil _____*Uniform sand*_____   Sample no. _____

Location _____

Length of specimen L ___*13.2*___ cm   Diameter of specimen D _____*6.35*_____ cm

Tested by _____   Date _____

Test No.	1	2	3
Average flow, Q (cm³)	305	375	395
Time of collection, t (s)	60	60	60
Temperature of water, T(°C)	25	25	25
Head difference, h (cm)	60	70	80
Diameter of specimen, D (cm)	6.35	6.35	6.35
Length of specimen, L (cm)	13.2	13.2	13.2
Area of specimen, $A = \dfrac{\pi}{4}D^2$ (cm²)	31.67	31.67	361.67
$k = \dfrac{QL}{Aht}$ (cm/s)	0.035	0.037	0.034

Average $k =$ ___*0.035*___ cm/s

$k_{20°C} = k_{T°C}\dfrac{\eta_{T°C}}{\eta_{20°C}} =$ ___*0.035 × 0.889* =___ ___*0.031*___ cm/s

11
Falling-Head Permeability Test in Sand

11.1 Introduction

The procedure for conducting the constant-head permeability tests in sand was discussed in the preceding chapter. The falling-head permeability test is another experimental procedure to determine the coefficient of permeability of sand.

11.2 Equipment

1. Falling-head permeameter
2. Balance sensitive to 0.1 g
3. Thermometer
4. Stopwatch

11.3 Falling-Head Permeameter

A schematic diagram of a falling-head permeameter is shown in Fig. 11.1. Its specimen tube is essentially the same as that used in the constant-head test. The top of the specimen tube is connected to a burette by plastic tubing. The specimen tube and the burette are held vertically by clamps from a stand. The bottom of the specimen tube is connected to a plastic funnel by a plastic tube. The funnel is held vertically by a clamp from another stand. A scale is also fixed vertically to this stand.

11.4 Procedure

1–9. Follow the same procedure for the preparation of the specimen as described in Chapter 10.

10. Supply water using a plastic tube from the water inlet to the burette. The water will flow from the burette to the specimen and then to the funnel. Check to see that there is no leak. Remove all air bubbles.

11. Allow the water to flow for some time in order to saturate the specimen. When the funnel is full, water will flow out of it into the sink.

12. Using the pinch cock, close the flow of water through the specimen. The pinch cock is located on the plastic pipe connecting the bottom of the specimen to the funnel.

13. Measure the head difference h_1 (cm) (see Fig. 11.1). (*Note: Do not* add any more water to the burette.)

14. Open the pinch cock. Water will flow through the burette to the specimen and then out of the funnel. Record time t with a stopwatch until the head difference is equal to h_2 (cm) (Fig. 11.1). Close the flow of water through the specimen using the pinch cock.

15. Determine the volume V_w of water (cm^3) that is drained from burette.

16. Add more water to the burette to make another run. Repeat Steps 13, 14, and 15. However, h_1 and h_2 should be changed for each run.

17. Record the temperature T of the water to the nearest degree (°C).

11.5 Calculations

The coefficient of permeability can be expressed by the relation

$$k = 2.303 \frac{aL}{At} \log \frac{h_1}{h_2} \tag{11.1}$$

where a is the inside cross-sectional area of the burette,[1]

$$a = \frac{V_w}{h_1 - h_2} \tag{11.2}$$

Therefore,

$$k = \frac{2.303 V_w L}{(h_1 - h_2)tA} \log \frac{h_1}{h_2} \tag{11.3}$$

where A is the area of the specimen.

As in Chapter 10,

$$k_{20°C} = k_{T°C} \frac{\eta_{T°C}}{\eta_{20°C}} \tag{11.4}$$

Sample calculations are shown in Tables 11.1 and 11.2.

---

[1] For an example of the derivation, see the References (Das, 2006).

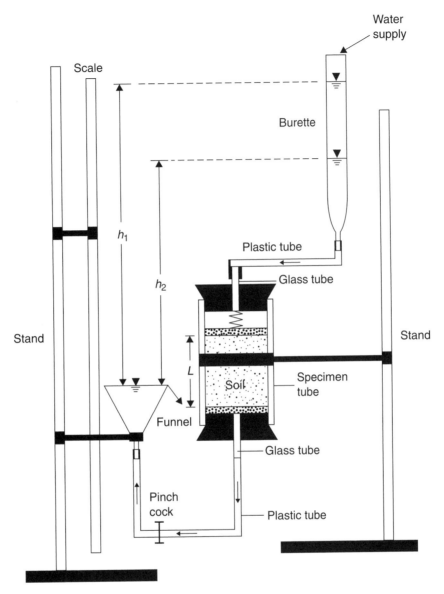

Figure 11.1. Schematic diagram of falling-head permeability test setup.

11.6 Discussion

The permeability test procedures described in Chapters 10 and 11 are using *rigid-wall permeameters*. In advanced geotechnical engineering courses, students will be exposed to flexible-wall permeameters, which is beyond the scope of this laboratory manual. However, generally speaking, flexible-wall permeability tests (constant and falling head) are performed in triaxial cells (see Chapter 18). According to ASTM test designation D-5084 the flexible-wall permeameter can be used for tests on soils having k less than about 10^{-4} cm/s. In a triaxial cell, interchangeable base pedestals and top caps permit the test of specimens with diameters from 40 to 150 mm. Drainage lines at the top and bottom of

Table 11.1. Determination of Void Ratio of Specimen—Falling-Head Permeability Test

Description of soil _____*Uniform sand*_____   Sample no. _____

Location _____

Length of specimen L _____13.2_____ cm   Diameter of specimen D _____6.35_____ cm

Tested by _____   Date _____

Volume of specimen, $V = \dfrac{\pi}{4}D^2L$ (cm^3)	418.03
Specific gravity of soil solids, G_s	2.66
Mass of specimen tube with fittings, M_1 (g)	238.4
Mass of tube with fittings and specimen, M_2 (g)	965.3
Dry density of specimen, $\rho_d = \dfrac{M_2 - M_1}{V}$ (g/cm^3)	1.74

Void ratio of specimen $e = \dfrac{G_s \rho_w}{\rho_d} - 1 =$ _____0.53_____

(*Note:* $\rho_w = 1$ g/cm^3)

the specimen facilitate the flushing of air bubbles from hydraulic lines and direct measurement of the pressure drop across the soil specimen using a differentially acting electrical pressure transducer. Separate pressure controls maintain the cell pressure and the pressures acting at the top and bottom of the soil specimen. Normally the specimen is back pressured prior to permeation to ensure full saturation.

The flexible-wall cell has several advantages. Undisturbed samples can be easily tested because minimal trimming is required and irregular surfaces on the specimen are easily accommodated. Back pressure is normally used, which helps to saturate the soil.

Flexible-wall cells have several disadvantages as well. The membranes used to confine the soil are normally made of latex, butyl, or neoprene rubber, which can be attacked and destroyed by certain chemicals. In order to maintain contact between the membrane and the soil specimen, the pressure in the cell liquid must be higher than the pore pressure in the specimen. In order to test with an elevated hydraulic gradient, the effective stress at one end of the specimen must be fairly large and the effective confining pressure cannot be less than the pressure drop across the specimen.

The coefficient of permeability k of fine-grained soils can also be determined from consolidation tests, which are discussed in Chapter 17.

Table 11.2. Determination of Coefficient of Permeability—Falling-Head Permeability Test

Description of soil _____ *Uniform sand* _____ Sample no. _____

Location _____

Length of specimen L _____ *13.2* _____ cm   Diameter of specimen D _____ *6.35* _____ cm

Tested by _____   Date _____

Test No.	1	2	3
Diameter of specimen, D (cm)	6.35	6.35	6.35
Length of specimen, L (cm)	13.2	13.2	13.2
Area of specimen, A (cm^2)	31.67	31.67	31.67
Beginning head difference, h_1 (cm)	85.0	76.0	65.0
Ending head difference, h_2 (cm)	24.0	20.0	20.0
Test duration, t (s)	15.4	15.3	14.4
Volume of water flow through specimen, V_w (cm^3)	64	58	47
$k = \dfrac{2.303 V_w L}{(h_1 - h_2)tA} \log \dfrac{h_1}{h_2}$ (cm/s)	0.036	0.038	0.036

Average $k =$ _____ *0.037* _____ cm/s

$k_{20°C} = k_{T°C} \dfrac{\eta_{T°C}}{\eta_{20°C}} =$ _____ *0.037 × 0.889* _____ = _____ *0.033* _____ cm/s

12
Standard Proctor Compaction Test

12.1 Introduction

For the construction of highways, airports, and other structures, it is often necessary to compact soil to improve its strength. Proctor (1933) developed a laboratory compaction test procedure to determine the maximum dry unit weight of compaction of soils which can be used for the specification of field compaction. This test is referred to as the *standard Proctor compaction test* and is based on the compaction of the soil fraction passing U.S. No. 4 sieve.

12.2 Equipment

1. Compaction mold
2. U.S. No. 4 sieve
3. Standard Proctor hammer of weight 5.5 lb (24.4 N)
4. Balance sensitive to 0.01 lb
5. Balance sensitive to 0.1 g
6. Large flat pan
7. Jack
8. Steel straightedge
9. Moisture cans
10. Drying oven
11. Plastic squeeze bottle with water

Figure 12.1 shows the equipment required for the compaction test with the exception of the jack, the balances, and the oven.

Figure 12.1. Equipment for Proctor compaction test.

12.3 Proctor Compaction Mold and Hammer

A schematic diagram of the Proctor compaction mold, which is 4 in. (101.6 mm) in diameter and 4.584 in. (116.43 mm) in height, is shown in Fig. 12.2(a). There is a base plate and an extension that can be attached to the top and bottom of the mold, respectively. The inside of the mold is 1/30 ft³ (943 cm³).

Figure 12.2(b) shows the schematic diagram of a standard Proctor hammer. The hammer can be lifted and dropped through a vertical distance of 12 in. (304.8 mm).

12.4 Procedure

1. Obtain about 10 lb (4.5 kg mass) of air-dry soil on which the compaction test is to be conducted. Break all the soil lumps.
2. Sieve the soil on a U.S. No. 4 sieve. Collect all of the minus 4 material in a large pan. This should be about 6 lb (2.7 kg mass) or more.
3. Add enough water to the minus 4 material and mix it in thoroughly to bring the moisture content up to about 5%.
4. Determine the weight of the Proctor mold + base plate (not the extension), W_1 (lb).
5. Now attach the extension to the top of the mold.
6. Pour the moist soil into the mold in *three* equal layers. Each layer should be compacted uniformly by the standard Proctor hammer 25 times before the next layer of loose soil is poured into the mold (see Fig. 12.3). (*Note:* The layers of loose soil that are being

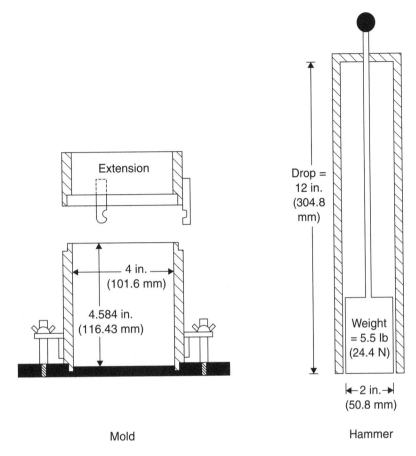

Figure 12.2. Standard Proctor mold and hammer.

poured into the mold should be such that, *at the end of the three-layer compaction,* the soil should extend *slightly above* the top of the rim of the compaction mold.)

7. Remove the top attachment from the mold. Be careful not to break off any of the compacted soil inside the mold while removing the top attachment.

8. Using a straightedge, trim the excess soil above the mold (Fig. 12.4). Now the top of the compacted soil will be even with the top of the mold.

9. Determine the weight of mold + base plate + compacted moist soil in mold, W_2 (lb).

10. Remove the base plate from the mold. Using a jack, extrude the compacted soil cylinder from the mold.

11. Take a moisture can and determine its mass, M_3 (g).

12. From the moist soil extruded in Step 10, collect a moisture sample in the moisture can (Step 11) and determine the mass of the can + moist soil, M_4 (g).

13. Place the moisture can with the moist soil in the oven to dry to a constant weight.

14. Break the rest of the compacted soil (to No. 4 size) by hand and mix it with the leftover moist soil in the pan. Add more water and mix it to raise the moisture content by about 2%.

Figure 12.3. Compaction of soil in Proctor mold.

15. Repeat Steps 6 through 12. In this process, the weight of the mold + base plate + moist soil (W_2) will first increase with the increase in moisture content and then decrease. Continue the test until at least two successive down readings are obtained.
16. The next day, determine the mass of the moisture cans + soil samples, M_5 (g) (from Step 13).

12.5 Calculations

Dry Unit Weight and Moisture Content at Compaction

The sample calculations for a standard Proctor compaction test are given in Table 12.1. Referring to Table 12.1,

Line 1. Weight of mold and base plate W_1, to be determined from test (Step 4).
Line 2. Weight of mold and base plate + moist compacted soil W_2, to be determined from test (Step 9).
Line 3. Weight of moist compacted soil, $= W_2 - W_1$ (line 2 − line 1).

Figure 12.4. Excess soil being trimmed (Step 8).

Line 4. Moist unit weight,

$$\gamma = \frac{\text{weight of compacted moist soil}}{\text{volume of mold}} = \frac{W_2 - W_1 \text{ (lb)}}{1/30\left(\text{ft}^3\right)}$$

$$= 30 \left(\text{lb/ft}^3\right) \times \text{line 3}$$

Line 6. Mass of moisture can M_3, to be determined from test (Step 11).
Line 7. Mass of moisture can + moist soil M_4, to be determined from test (Step 12).
Line 8. Mass of moisture can + dry soil M_5, to be determined from test (Step 16).
Line 9. Compaction moisture content M_5,

$$w\ (\%) = \frac{M_4 - M_5}{M_5 - M_3} \times 100$$

Line 10. Dry unit weight,

$$\gamma_d = \frac{\gamma}{1 + (w(\%)/100)} = \frac{\text{line 4}}{1 + (\text{line 9}/100)}$$

Table 12.1. Determination of Dry Unit Weight—Standard Proctor
Compaction Test

Description of soil _____*Light brown clayey silt*_____ Sample no. ___*2*___

Location _____

Volume of mold _____*1/30*_____ ft³ Weight of hammer _____*5.5*_____ lb

Number of blows/layer _____*25*_____ Number of layers _____*3*_____

Tested by _____ Date _____

Test	1	2	3	4	5	6
1. Weight of mold and base plate, W_1 (lb)	10.35	10.35	10.35	10.35	10.35	10.35
2. Weight of mold and base plate + moist soil, W_2 (lb)	14.19	14.41	14.53	14.63	14.51	14.47
3. Weight of moist soil, $W_2 - W_1$ (lb)	3.84	4.06	4.18	4.28	4.16	4.12
4. Moist unit weight, $\gamma = \dfrac{W_2 - W_1}{1/30}$ (lb/ft³)	115.2	121.8	125.4	128.4	124.8	123.8
5. Moisture can number	202	212	222	242	206	504
6. Mass of moisture can, M_3 (g)	54.0	53.3	53.3	54.0	54.8	40.8
7. Mass of can + moist soil, M_4 (g)	253.0	354.0	439.0	490.0	422.8	243.0
8. Mass of can + dry soil, M_5 (g)	237.0	326.0	401.0	441.5	374.7	211.1
9. Moisture content, $w\,(\%) = \dfrac{M_4 - M_5}{M_5 - M_3} \times 100$	8.7	10.3	10.9	12.5	15.0	18.8
10. Dry unit weight of compaction, γ_d (lb/ft³) $= \dfrac{\gamma}{1+(w\,(\%)/100)}$	106.0	110.4	113.0	114.1	108.5	104.2

Table 12.2. Zero-Air-Void Unit Weight—Standard Proctor Compaction Test

Description of soil _____ _Light brown clayey silt_ _____ Sample no. __2__

Location _____

Tested by _____ Date _____

Specific Gravity of Soil Solids G_s	Assumed Moisture Content w (%)	Unit Weight of Water γ_w (lb/ft^3)	γ_{zav}^* (lb/ft^3)
2.68	10	62.4	131.9
2.68	12	62.4	126.5
2.68	14	62.4	121.6
2.68	16	62.4	117.0
2.68	18	62.4	112.8
2.68	20	62.4	108.7

*Equation (12.1).

Zero-Air-Void Unit Weight

The maximum theoretical dry unit weight of a compacted soil at a given moisture content will occur when there is no air left in the void spaces of the compacted soil. This can be given by

$$\gamma_{d(\text{theory}-\text{max})} = \gamma_{zav} = \frac{\gamma_w}{(w\,(\%)\,/100) + (1/G_S)} \tag{12.1}$$

where γ_{zav} = zero-air-void unit weight
γ_w = unit weight of water
w = moisture content
G_s = specific gravity of soil solids

Since the values of γ_w and G_s will be known, several values of w (%) can be assumed and γ_{zav} can be calculated. Table 12.2 shows the calculations for γ_{zav} for the soil tested and reported in Table 12.1.

12.6 Graphs

Plot a graph showing γ_d (line 10, Table 12.1) versus w (%) (line 9, Table 12.1) and determine the *maximum dry unit weight of compaction* $\gamma_{d(\text{max})}$. Also determine the *optimum moisture*

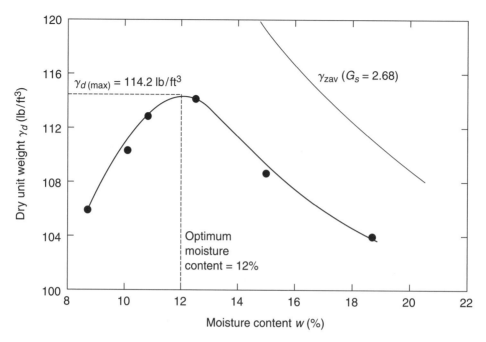

Figure 12.5. Plots of γ_d vs. w (%) and γ_{zav} vs. w (%) for test results reported in Tables 12.1 and 12.2.

content w_{opt}, which is the moisture content corresponding to $\gamma_{d(max)}$. On the same graph, plot γ_{zav} versus w (%). (*Note:* For a given soil, *no portion* of the experiment curve of γ_d versus w(%) should plot to the *right* of the zero-air-void curve.)

Figure 12.5 shows the results of the calculations made in Tables 12.1 and 12.2.

12.7 General Comments

A. Units

1. The test presented in this chapter uses English units to express γ or γ_d. If there is a need for conversion to SI units,

$$\gamma \text{ or } \gamma_d \left(\text{kN/m}^3 \right) = 0.15706 \, \gamma \text{ or } \gamma_d \left(\text{lb/ft}^3 \right)$$

2. In several instances, referring to Steps 4 and 9 (Section 12.4, Procedure), the mass is determined in kg. In that case,
   - Mass of Proctor mold + base plate, M_1 (kg)
     (compare with Procedure, Step 4)
   - Mass of mold + base plate + compacted moist soil in mold, M_2 (kg)
     (compare with Procedure, Step 9)

- Moist density,

$$\rho\,(\text{kg/m}^3) = \frac{M_2 - M_1\,(\text{kg})}{943 \times 10^{-6}\,(\text{m}^3)}$$

(compare with line 4 in Table 12.1)
- Dry density,

$$\rho_d\,(\text{kg/m}^3) = \frac{\rho\,(\text{kg/m}^3)}{1 + w\,(\%)/100}$$

(compare with line 10 in Table 12.1)
- Zero-air-void density,

$$\rho_{d\,(\text{theory} - \text{max})} = \rho_{\text{zav}}\,(\text{kg/m}^3) = \frac{\rho_w}{(w\,(\%)\,/100) + (1/G_s)}$$

where ρ_w is the density of water, $= 1000\ \text{kg/m}^3$ [compare with Eq. (12.1)].

B. Relative Compaction

In most specifications for earth work it is required to achieve a compacted field dry unit weight of 90–95% of the maximum dry unit weight obtained in the laboratory. This is sometimes referred to as relative compaction R, or

$$R\,(\%) = \frac{\gamma_{d\,(\text{field})}}{\gamma_{d\,(\text{max}-\text{lab})}} \times 100 \qquad (12.2)$$

For granular soils it can be shown that

$$R\,(\%) = \frac{R_0}{1 - D_r\,(1 - R_0)} \times 100 \qquad (12.3)$$

where D_r is the relative density of compaction and

$$R_0 = \frac{\gamma_{d\,(\text{max})}}{\gamma_{d\,(\text{min})}} \qquad (12.4)$$

Compaction of cohesive soils will influence their structure, coefficient of permeability, one-dimensional compressibility, and strength. For further discussion on this topic, refer to Das (2006).

Summary of ASTM Proctor Test Specifications

In this chapter the laboratory test outlines given for compaction tests use the following:

Volume of mold $= 1/30\ \text{ft}^3$ (943 cm^3)
Number of blows $= 25$

Table 12.3. Summary of Standard Proctor Compaction Test Specifications (ASTM D-698)

Description	Method A*	Method B†	Method C‡
Mold			
Volume	1/30 ft^3 (943 cm^3)	1/30 ft^3 (943 cm^3)	1/13.33 ft^3 (2124 cm^3)
Height	4.584 in. (116.43 mm)	4.584 in. (116.43 mm)	4.584 in. (116.43 mm)
Diameter	4 in. (101.6 mm)	4 in. (101.6 mm)	6 in. (152.4 mm)
Weight of hammer	5.5 lb (24.4 N)	5.5 lb (24.4 N)	5.5 lb (24.4 N)
Height of hammer drop	12 in. (304.8 mm)	12 in. (304.8 mm)	12 in. (304.8 mm)
Number of layers of soil	3	3	3
Number of blows per layer	25	25	56
Test on soil fraction passing sieve	No. 4 (4.75 mm)	3/8 in. (9.5 mm)	3/4 in. (19.0 mm)

* May be used if 20% or less by mass of material is retained on No. 4(4.75-mm) U.S. sieve.

† May be used if more than 20% by mass of material is retained on No. 4 (4.75-mm) U.S. sieve and 20% or less by mass of material is retained on 3/8-in. (9.5-mm) U.S. sieve.

‡ May be used if more than 20% by mass of material is retained on 3/8-in. (9.5-mm) U.S. sieve and less than 30% by mass of material is retained on 3/4-in. (19.0-mm) U.S. sieve.

These values are generally used for fine-grained soils that pass through No. 4 U.S. sieve. However, ASTM has three different methods for the standard Proctor compaction test that reflect the size of the mold, the number of blows per layer, and the maximum particle size in a soil used for testing. Summaries of these methods are given in Table 12.3.

13
Modified Proctor Compaction Test

13.1 Introduction

In the preceding chapter we have seen that water generally acts as a lubricant between solid particles during the soil compaction process. Because of this, in the initial stages of compaction, the dry unit weight of compaction increases. However, another factor that will control the dry unit weight of compaction of a soil at a given moisture content is the energy of compaction. For the standard Proctor compaction test, the energy of compaction can be given by

$$\frac{(3 \text{ layers})(25 \text{ blows/layer})(5.5 \text{ lb})(1 \text{ft/blow})}{1/30 \text{ ft}^3} = 12,375 \frac{\text{ft} \cdot \text{lb}}{\text{ft}^3} \times (\approx 600 \text{ kN} \cdot \text{m/m}^3)$$

The modified Proctor compaction test is a standard test procedure for the compaction of soil using a higher energy of compaction. In this test, the compaction energy is equal to

$$56,250 \frac{\text{ft} \cdot \text{lb}}{\text{ft}^3} (\approx 2700 \text{ kN} \cdot \text{m/m}^3)$$

13.2 Equipment

The equipment required for the modified Proctor compaction test is the same as in Chapter 12, with the exception of the standard Proctor hammer (item 3). The hammer used for this test weighs 10 lb (44.5 N) and drops through a vertical distance of 18 in. (457.2 mm). Figure 13.1 shows the standard and modified Proctor test hammers side by side.

Figure 13.1. Comparison of standard and modified Proctor compaction hammers. (*Note:* Hammer at right is for modified Proctor compaction test.)

The compaction mold used in this test is the same as that described in Chapter 12 [i.e., volume = $1/30$ ft^3 (943 cm^3)].

13.3 Procedure

The procedure is the same as that described in Chapter 12, except for Step 6. The moist soil has to be poured into the mold in five equal layers. Each layer has to be compacted by the modified Proctor hammer with 25 blows per layer.

13.4 Calculations, Graphs, and Zero-Air-Void Curve

Same as in Chapter 12.

13.5 General Comments

1. The modified Proctor compaction test results for the same soil as reported in Tables 12.1 and 12.2 and Fig. 12.5 are shown in Fig. 13.2. A comparison of γ_d versus w (%) curves obtained from standard and modified Proctor compaction tests shows that
   a. The maximum dry unit weight of compaction increases with the increase in the compacting energy, and
   b. The optimum moisture content decreases with the increase in the energy of compaction.
2. As reported in Chapter 12, there are three different methods suggested by ASTM for this test. They are shown in Table 13.1.
3. More recently, Gurtug and Sridharan (2004) proposed approximate correlations for optimum moisture content and maximum dry unit weight with the plastic limit (*PL*) of cohesive soils. These correlations can be expressed as

$$w_{\text{opt}}(\%) = [1.95 - 0.38 \log (CE)] \, PL \tag{13.1}$$

$$\gamma_{d\,(\text{max})} \left(\text{kN/m}^3\right) = 22.68 e^{-0.0183 w_{\text{opt}}(\%)} \tag{13.2}$$

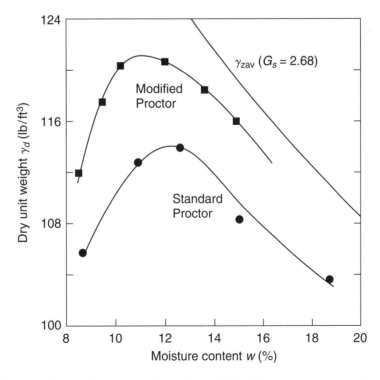

Figure 13.2. Comparison of standard and modified Proctor compaction test results for soil reported in Tables 12.1 and 12.2.

Table 13.1. Summary of Standard Proctor Compaction Test Specifications (ASTM D-1556)

Description	Method A*	Method B†	Method C‡
Mold			
Volume	$1/30$ ft^3 (943 cm^3)	$1/30$ ft^3 (943 cm^3)	$1/13.33$ ft^3 (2124 cm^3)
Height	4.584 in. (116.43 mm)	4.584 in. (116.43 mm)	4.584 in. (116.43 mm)
Diameter	4 in. (101.6 mm)	4 in. (101.6 mm)	6 in. (152.4 mm)
Weight of hammer	10 lb (44.5 N)	10 lb (44.5 N)	10 lb (44.5 N)
Height of hammer drop	18 in. (457.2 mm)	18 in. (457.2 mm)	18 in. (457.2 mm)
Number of layers of soil	5	5	5
Number of blows per layer	25	25	56
Test on soil fraction passing sieve	No. 4 (4.75 mm)	3/8 in. (9.5 mm)	3/4 in. (19.0 mm)

* May be used if 20% or less by mass of material is retained on No. 4 (4.75-mm) U.S. sieve.

† May be used if more than 20% by mass of material is retained on No. 4 (4.75-mm) U.S. sieve and 20% or less by mass of material is retained on 3/8-in. (9.5-mm) U.S. sieve.

‡ May be used if more than 20% by mass of material is retained on 3/8-in. (9.5-mm) U.S. sieve and less than 30% by mass of material is retained on 3/4-in (19.0-mm) U.S. sieve.

where PL = plastic limit (%)

CE = compaction energy (kN · m/m^3)

For standard Proctor tests, $CE = 600$ kN · m/m^3. Thus,

$$w_{\text{opt}} (\%) = [1.95 - 0.38 \log (600)] PL = 0.894\, PL \tag{13.3}$$

$$\gamma_{d(\text{max})} (\text{kN/m}^3) = 22.68 e^{-0.0164\, PL} \tag{13.4}$$

or

$$\gamma_{d(\text{max})} (\text{lb/ft}^3) = 144.26 e^{-0.0164\, PL} \tag{13.5}$$

For modified Proctor tests, $CE = 2700 \text{ kN} \cdot \text{m/m}^3$. So,

$$w_{\text{opt}} (\%) = [1.95 - 0.38 \log (2700)] PL = 0.646 \, PL \qquad (13.6)$$

$$\gamma_{d(\text{max})} \left(\text{kN/m}^3 \right) = 22.68e^{-0.0118 \, PL} \qquad (13.7)$$

$$\gamma_{d(\text{max})} \left(\text{lb/ft}^3 \right) = 144.26e^{-0.0118 \, PL} \qquad (13.8)$$

It should be kept in mind that these correlations are not a substitution for the actual laboratory tests.

14

Determination of Field Unit Weight of Compaction by Sand Cone Method

14.1 Introduction

In the field during soil compaction it is sometimes necessary to check the compacted dry unit weight of soil and compare it with the specifications drawn up for the construction. One of the simplest methods of determining the field unit weight of compaction is by the sand cone method, which will be described in this chapter.

14.2 Equipment

1. Sand cone apparatus consisting of a one-gallon glass or plastic bottle with a metal cone attached to it
2. Base plate
3. One-gallon can with cap
4. Tools to dig a small hole in the field
5. Balance with minimum readability of 0.01 lb
6. 20-30 Ottawa sand
7. Proctor compaction mold without attached extension
8. Steel straightedge

Figure 14.1 shows the assembly of the equipment necessary for the determination of the field unit weight. Figure 14.2 is a schematic diagram showing the dimensions of the metal cone (see item 1).

Figure 14.1. Assembly of equipment necessary for determination of field unit weight of compaction.

Bottle attached here

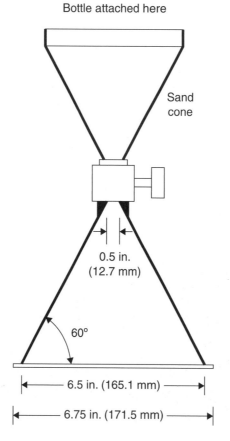

Sand cone

0.5 in. (12.7 mm)

60°

6.5 in. (165.1 mm)

6.75 in. (171.5 mm)

Figure 14.2. Dimensions of metal cone.

14.3 Procedure—Laboratory Work

Determination of Dry Unit Weight of 20-30 Ottawa Sand

1. Determine the weight of the Proctor compaction mold, W_1.
2. Using a spoon, fill the compaction mold with 20-30 Ottawa sand. Avoid any vibration or other means of compaction of the sand poured into the mold. When the mold is full, strike off the top of the mold with the steel straightedge. Determine the weight of the mold and sand, W_2.

Calibration of Cone

3. We need to determine the weight of the Ottawa sand that is required to fill the cone. This can be done by filling the one-gallon bottle with Ottawa sand. Determine the weight of the bottle + cone + sand, W_3. Close the valve of the cone, which is attached to the bottle. Place the base plate on a flat surface. Turn the bottle with the cone attached to it upside down and place the open mouth of the cone in the center hole of the base plate (Fig. 14.3). Open the cone valve. Sand will flow out of the bottle and gradually fill the cone. When the cone is filled with sand, the flow of sand from the bottle will stop.

Figure 14.3. Calibration of sand cone.

Close the cone valve. Remove the bottle and cone combination from the base plate and determine its weight, W_4.

Preparation for Field Work

4. Determine the weight of the gallon can without the cap, W_5.
5. Fill the one-gallon bottle (with the sand cone attached to it) with sand. Close the valve of the cone. Determine the weight of the bottle + cone + sand, W_6.

14.4 Procedure—Field Work

6. Now proceed to the field with the bottle and the cone attached to it (filled with Ottawa sand—Step 5), the base plate, the digging tools, and the one-gallon can with its cap.
7. Place the base plate on a level ground in the field. Under the center hole of the base plate, dig a hole in the ground using the digging tools. The volume of the hole should be smaller than the volume of the sand in the bottle minus the volume of the cone.
8. Remove *all* the loose soil from the hole and put it in the gallon can. Close the cap tightly so as not to lose any moisture. Be careful not to move the base plate.
9. Turn the gallon bottle filled with sand, with cone attached to it, upside down and place it on the center of the base plate. Open the valve of the cone. Sand will flow from the bottle to fill the hole in the ground and the cone (Fig. 14.4). When the flow of sand from the bottle stops, close the valve of the cone and remove it.

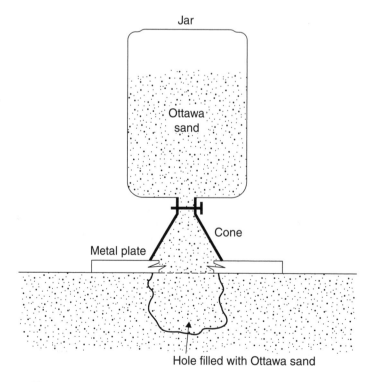

Figure 14.4. Ottawa sand filling hole in ground.

10. Bring all the equipment back to the laboratory. Determine the weight of the gallon can + moist soil from the field (without the cap), W_7. Also determine the weight of the bottle + can + sand after use, W_8.
11. Put the gallon can with the moist soil in the oven to dry to a constant weight. Determine the weight of the can without the cap + oven-dry soil, W_9.

14.5 Calculations

A sample calculation to determine the dry unit weight of field compaction by the sand cone method is given in Table 14.1. With reference to the table, the following calculations are required.

1. Calculate the dry unit weight of sand (line 4),

$$\gamma_{d(\text{sand})} = \frac{W_2 - W_1}{V_1} = \frac{\text{line } 2 - \text{line } 1}{\text{line } 3} \tag{14.1}$$

where V_1 is the volume of the Proctor mold.
2. Calculate the weight of sand to fill the cone (line 7),

$$W_c = W_4 - W_3 = \text{line } 6 - \text{line } 5 \tag{14.2}$$

3. Calculate the volume of the hole in the field (line 10),

$$V_2 = \frac{W_6 - W_8 - W_c}{\gamma_{d(\text{sand})}} = \frac{\text{line } 8 - \text{line } 9 - \text{line } 7}{\text{line } 4} \tag{14.3}$$

4. Calculate the moist field unit weight (line 14),

$$\gamma = \frac{W_7 - W_5}{V_2} - \frac{\text{line } 12 - \text{line } 11}{\text{line } 10} \tag{14.4}$$

5. Calculate the moisture content in the field (line 15),

$$w\,(\%) = \frac{W_7 - W_9}{W_9 - W_5} \times 100 = \frac{\text{line } 12 - \text{line } 13}{\text{line } 13 - \text{line } 11} \times 100 \tag{14.5}$$

6. Calculate the dry unit weight in the field (line 16),

$$\gamma_d = \frac{\gamma}{1 + (w\,(\%)/100)} = \frac{\text{line } 14}{1 + (\text{line } 15/100)} \tag{14.6}$$

14.6 General Comments

There are at least two other methods to determine the field unit weight of compaction. They are the *rubber balloon method* (ASTM D-2167) and use of the *nuclear density meter*. The procedure for the rubber balloon method is similar to the sand cone method in that a test hole is made, the moist weight of the soil is removed from the hole, and its moisture

Table 14.1.   Field Unit Weight—Sand Cone Method

Location _____ *Madeira Canyon/Provence* _____

Tested by _____   Date _____

Item	Quantity
Calibration of Unit Weight of Ottawa Sand	
1. Weight of Proctor mold, W_1	10.35 lb
2. Weight of Proctor mold + sand, W_2	13.66 lb
3. Volume of mold, V_1	1/30 ft^3
4. Dry unit weight, $\gamma_{d(\text{sand})} = \dfrac{W_2 - W_1}{V_1}$	99.3 lb/ft^3
Calibration Cone	
5. Weight of bottle + cone + sand (before use), W_3	15.17 lb
6. Weight of bottle + cone + sand (after use), W_4	14.09 lb
7. Weight of sand to fill cone, $W_c = W_4 - W_3$	1.08 lb
Results from Field Tests	
8. Weight of bottle + cone + sand (before use), W_6	15.42 lb
9. Weight of bottle + cone + sand (after use), W_8	11.74 lb
10. Volume of hole, $V_2 = \dfrac{W_6 - W_8 - W_c}{\gamma_{d(\text{sand})}}$	0.0262 ft^3
11. Weight of gallon can, W_5	0.82 lb
12. Weight of gallon can + moist soil, W_7	3.92 lb
13. Weight of gallon can + dry soil, W_9	3.65 lb
14. Moist unit weight of soil in field, $\gamma = \dfrac{W_7 - W_5}{V_2}$	118.32 lb/ft^3
15. Moisture content in field, $w\,(\%) = \dfrac{W_7 - W_9}{W_9 - W_5} \times 100$	9.54%
16. Dry unit weight in field, $\gamma_{d(\text{sand})} = \dfrac{\gamma}{1 + (w\,(\%)/100)}$	108.11 lb/ft^3

content is determined. However, the volume of the hole is determined by introducing into it a rubber balloon filled with water from a calibrated vessel, from which the volume can be read directly.

Nuclear density meters are now used in some large projects to determine the compacted dry unit weight of a soil (ASTM D-2922). The density meters operate either in drilled holes or from the ground surface. The instrument measures the weight of the wet soil per unit volume and also the weight of the water present in a volume of soil. The dry unit weight of compacted soil can be determined by subtracting the weight of the water from the moist unit weight of the soil.

15
Direct Shear Test on Sand

15.1 Introduction

The shear strength s of a granular soil may be expressed by the equation

$$s = \sigma' \tan \phi' \tag{15.1}$$

where σ' = effective normal stress
ϕ' = angle of friction of soil

The angle of friction ϕ' is a function of the relative density of compaction of sand, grain size, shape, and distribution in a given soil mass. For a given sand, an increase in the void ratio (i.e., a decrease in the relative density of compaction) will result in a decrease of the magnitude of ϕ'. However, for a given void ratio, an increase in the angularity of the soil particles will give a higher value of the soil friction angle. The general range of the drained angle of friction of sand for various values of relative density is given in Table 15.1.

15.2 Equipment

1. Direct shear test machine (strain controlled)
2. Balance sensitive to 0.1 g
3. Large porcelain evaporating dish
4. Tamper (for compacting sand in the direct shear box)
5. Spoon

Figure 15.1 shows a direct shear test machine. It consists primarily of a direct shear box, which is split into two halves (top and bottom) and which holds the soil specimen; a proving ring to measure the horizontal load applied to the specimen; two dial gauges (one horizontal and one vertical) to measure the deformation of the soil during the test; and a yoke by which a vertical load can be applied to the soil specimen. A horizontal load is applied to the top half of the shear box by a motor and gear arrangement. In a strain-controlled unit, the rate

Table 15.1. General Range of ϕ'

State of Sand	Relative Density (%)	ϕ' (deg)
Round grained sand		
Loose	0–50	28–32
Medium	50–70	30–35
Dense	70–100	35–40
Angular grained sand		
Loose	0–50	30–36
Medium	50–70	34–40
Dense	70–100	40–45

Figure 15.1. Strain-controlled direct shear test machine.

110

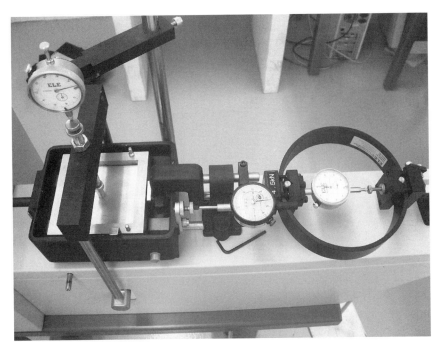

Figure 15.2. Shear box, proving ring, and horizontal and vertical dial gauges of strain-controlled direct shear test machine.

of movement of the top half of the shear box can be controlled. Figure 15.2 shows a top view of a direct shear machine, including the direct shear box, horizontal and vertical dial gauges, and the proving ring.

Figure 15.3 shows a schematic diagram of the shear box, which is split into two halves—top and bottom. These top and bottom halves can be held together by two vertical pins. There is a loading head, which can be slipped from the top of the shear box to rest on the soil specimen inside the box. There are also three vertical screws and two horizontal screws on the top half of the shear box.

15.3 Procedure

1. Remove the shear box assembly. Back off the three vertical and two horizontal screws. Remove the loading head. Insert the two vertical pins to keep the two halves of the shear box together.
2. Determine the mass of some dry sand in a large porcelain dish, M_1. Fill the shear box with sand in small layers. A tamper may be used to compact the sand layers. The top of the compacted specimen should be about 1/4 in. (6.4 mm) below the top of the shear box. Level the surface of the sand specimen. Determine the mass of the porcelain dish and dry sand remaining after compaction, M_2.
3. Determine the dimensions of the soil specimen (i.e., length L, width B, and height H).

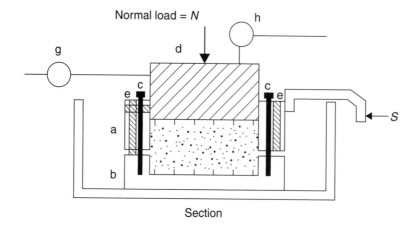

Section

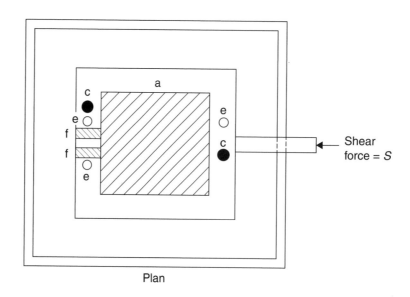

Plan

LEGEND

a – Top half of shear box
b – Bottom half of shear box
c – Vertical pins
d – Loading head
e – Vertical screw
f – Horizontal screw
g – Horizontal dial gauge
h – Vertical dial gauge

Figure 15.3. Schematic diagram of direct shear test box.

4. Slip the loading head down from the top of the shear box to rest on the soil specimen.

5. Put the shear box assembly in place in the direct shear machine.

6. Apply the desired normal load N on the specimen. This can be done by hanging dead weights to the vertical load yoke. The top crossbars will rest on the loading head of the specimen which, in turn, rests on the soil specimen. *Note:* In some equipment, the weights of the hanger, the loading head, and the top half of the shear box can be tared. In some other equipment, if taring is not possible, the normal load should be calculated as $N = $ load hanger + weight of yoke + weight of loading head + weight of top half of the shear box.

7. Remove the two vertical pins (which were inserted in Step 1 to keep the two halves of the shear box together).

8. Advance the three vertical screws that are located on the sidewalls of the top half of the shear box. This is done to separate the two halves of the box. The space between the two halves of the box should be slightly larger than the largest grain size of the soil specimen (by visual observation).

9. Set the loading head by tightening the two horizontal screws located at the top half of the shear box. Now back off the three vertical screws. After doing this, there will be no connection between the two halves of the shear box except the soil.

10. Attach the horizontal and vertical dial gauges (0.001 in./small div) to the shear box to measure the displacement during the test.

11. Apply horizontal load S to the top half of the shear box. The rate of shear displacement should be between 0.1 and 0.02 in./min (2.54–0.51 mm/min). For every tenth small division displacement in the horizontal dial gauge, record the readings of the vertical dial gauge and the proving ring gauge (which measures horizontal load S). Continue this until after either
    a. The proving ring dial gauge reading reaches a maximum and then falls, or
    b. The proving ring dial gauge reading reaches a maximum and then remains constant.

12. Repeat the test (Steps 1 to 11) at least two more times. For each test, the dry unit weight of compaction of the sand specimen should be the same as that of the first specimen (Step 2).

15.4 Calculations

Sample calculations of a direct shear test in sand for one normal load N are shown in Tables 15.2 and 15.3. Referring to Table 15.2:

1. Determine the dry density ρ_d of the specimen as

$$\rho_d = \frac{M_1 - M_2}{LBH} \tag{15.2}$$

Table 15.2. Void Ratio Calculation—Direct Shear Test on Sand

Description of soil _____ *Uniform sand* _____ Sample no. ___2___

Location _____ *Argonaut Circle* _____

Tested by _____ Date _____

Item	Quantity
1. Specimen length, L (in.)	2
2. Specimen width, B (in.)	2
3. Specimen height, H (in.)	1.31
4. Mass of porcelain dish + dry sand (before use), M_1 (g)	540.3
5. Mass of porcelain dish + dry sand (after use), M_2 (g)	397.2
6. Dry unit weight of specimen, $\gamma_d \, (\text{lb/ft}^3) = \dfrac{M_1 - M_2 (\text{g})}{LBH \, (\text{in.}^3)} \times 3.808$	104.0
7. Specific gravity of soil solids, G_s	2.66
8. Void ratio, $e = \dfrac{G_s \gamma_w}{\gamma_d} - 1$ (*Note:* $\gamma_w = 62.4 \, \text{lb/ft}^3$; γ_d is in lb/ft^3)	0.596

2. Determine the dry unit weight of the specimen (line 6). If M_1 and M_2 are in grams and $L, B,$ and H are in inches,

$$\gamma_d \left(\text{lb/ft}^3\right) = \frac{M_1 - M_2}{LBH} \times 3.808 \qquad (15.3)$$

3. Calculate the void ratio of the specimen (line 8),

$$e = \frac{G_s \gamma_w}{\gamma_d} - 1 \qquad (15.4)$$

where G_s is the specific gravity of soil solids, γ_w the unit weight of water (62.4 lb/ft^3), and γ_d is in lb/ft^3. Now refer to Table 15.3.

Table 15.3. Stress and Displacement Calculation—Direct Shear Test on Sand

Description of soil _____ *Uniform sand* _____ Sample no. ___ *2* ___

Location _____ *Argonaut Circle* _____

Normal load N _____ *56* _____ lb   Void ratio e _____ *0.596* _____

Tested by _____ Date _____

Normal Stress σ' (lb/in.²) (1)	Horizontal Displacement (in.) (2)	Vertical Displacement* (in.) (3)	No. of div. in Proving Ring Dial Gauge (4)	Proving Ring Calibration Factor (lb/div.) (5)	Shear Force S (lb) (6)	Shear Stress τ (lb/in.²) (7)
14	0	0	0	0.31	0	0
14	0.01	+0.001	45	0.31	13.95	3.49
14	0.02	+0.002	76	0.31	23.56	5.89
14	0.03	+0.004	95	0.31	29.76	7.44
14	0.04	+0.006	112	0.31	34.72	8.68
14	0.05	+0.008	124	0.31	38.44	9.61
14	0.06	+0.009	129	0.31	39.99	10.00
14	0.07	+0.010	125	0.31	38.75	9.69
14	0.08	+0.010	119	0.31	36.89	9.22
14	0.09	+0.009	114	0.31	35.34	8.84
14	0.10	+0.008	109	0.31	33.79	8.45
14	0.11	+0.008	108	0.31	33.48	8.37
14	0.12	+0.008	105	0.31	32.55	8.14

* Plus (+) sign means expansion.

4. Calculate the normal stress σ' on the specimen (column 1),

$$\sigma' \text{ (lb/in.}^2) = \frac{\text{normal load } N \text{ (lb)}}{L \text{ (in.)} \times B \text{ (in.)}} \tag{15.5}$$

5. The horizontal, vertical, and proving ring dial gauge readings are obtained from the test (columns 2, 3, and 4).
6. For any given set of horizontal and vertical dial gauge readings, calculate the shear force (column 6),

$$S = [\text{no. of divisions in proving ring dial gauge (column 4)}]$$
$$\times [\text{proving ring calibration factor (column 5)}] \tag{15.6}$$

7. Calculate the shear stress τ (column 7),

$$\tau \text{ (lb/in.}^2) = \frac{\text{shear force } S}{\text{area of specimen}} = \frac{S \text{ (lb)}}{L \text{ (in.)} \times B \text{ (in.)}} \tag{15.7}$$

Note: A separate data sheet has to be used for each test (i.e., for each normal stress σ').

15.5 Graphs

1. For each normal stress, plot a graph of τ (column 7) versus horizontal displacement (column 2), as shown in Fig. 15.4 for the results obtained from Table 15.2. On the bottom of the same graph paper, using the same horizontal scale, plot a graph of vertical displacement (column 3) versus horizontal displacement (column 2). There will be at least three such plottings (one for each value of σ'). Determine the shear stress at failure s from each τ versus horizontal displacement graph (as shown in Fig. 15.4). Note that the shear stress at failure is the shear strength.
2. Plot a graph of shear strength s versus normal stress σ'. This graph will be a straight line passing through the origin. Figure 15.5 shows such a plot for the sand reported in Tables 15.2 and 15.3. The angle of friction of the soil can be determined from the slope of the straight-line plot of s versus σ' as

$$\phi' = \tan^{-1}\left(\frac{s}{\sigma'}\right) \tag{15.8}$$

15.6 General Comments

The nature of the plots of shear stress and vertical displacement versus horizontal displacement shown in Fig. 15.4 may vary depending on the denseness of the sand. Figure 15.6 shows typical plots of shear stress and vertical displacement of specimens against horizontal

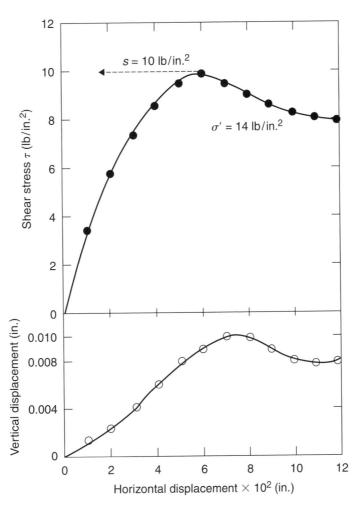

Figure 15.4. Plots of shear stress and vertical displacement vs. horizontal displacement for direct shear test reported in Tables 15.1 and 15.2.

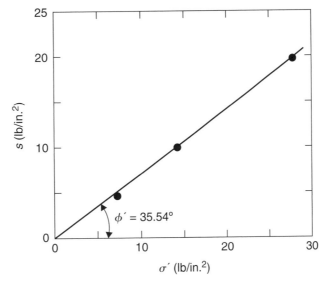

Figure 15.5. Plot of s vs. σ' for sand reported in Tables 15.1 and 15.2. (*Note:* results for tests with $\sigma' = 7$ 1b/in.^{2} and 28 1b/in.^{2} are not shown in Table 15.2.)

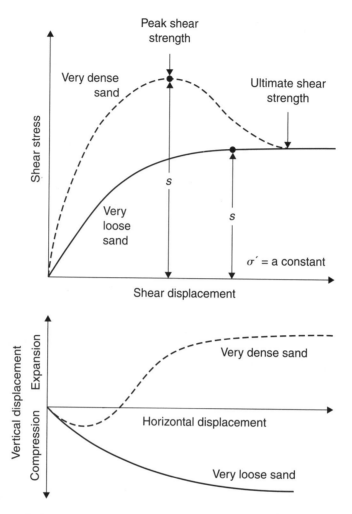

Figure 15.6. Plots of shear stress and vertical displacement of specimens vs. horizontal displacement for very loose and very dense sand.

displacement for dry very loose and very dense sands. The following generalizations can be developed:

1. In very loose sand, the resisting shear stress increases with shear displacement until a failure shear stress *s* (which is the shear strength) is reached. After that, the shear resistance remains approximately constant for any further increase in the shear displacement.
2. In very dense sand, the resisting shear stress increases with shear displacement until it reaches a failure stress *s*, called the *peak shear strength*. After failure stress is attained, the resisting shear stress gradually decreases as shear displacement increases until it finally reaches a constant value, called the *ultimate shear strength*.

16
Unconfined Compression Test

16.1 Introduction

The shear strength of a soil can be given by the Mohr–Coulomb failure criterion as

$$s = c' + \sigma' \tan \phi' \qquad (16.1)$$

where s = shear strength
c' = cohesion
σ' = effective normal stress
ϕ' = angle of friction

For undrained tests of saturated clayey soils ($\phi = 0$),

$$s = c_u \qquad (16.2)$$

where c_u is the undrained cohesion, or undrained shear strength.

The unconfined compression test is a quick method of determining the value of c_u for a clayey soil. The unconfined strength is given by the relation

$$c_u = \frac{q_u}{2} \qquad (16.3)$$

where q_u is the unconfined compression strength. For further discussion see any soil mechanics text, e.g., Das (2006).

The unconfined compression strength is determined by applying an axial stress to a cylindrical soil specimen with no confining pressure and observing the axial strains corresponding to various stress levels. The stress at which failure in the soil specimen occurs is referred to as the *unconfined compression strength* (Fig. 16.1). For *saturated* clay specimens,

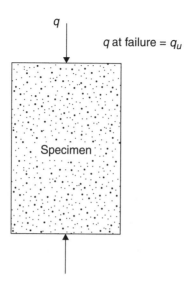

Figure 16.1. Definition of unconfined compression strength.

the unconfined compression strength decreases with the increase in moisture content. For *unsaturated* soils, with the dry unit weight remaining constant, the unconfined compression strength decreases with the increase in the degree of saturation.

16.2 Equipment

1. Unconfined compression testing device
2. Specimen trimmer and accessories (if undisturbed field specimen is used)
3. Harvard miniature compaction device and accessories (if specimen is to be molded for classroom work)
4. Scale
5. Balance sensitive to 0.01 g
6. Oven
7. Porcelain evaporating dish

16.3 Unconfined Compression Test Machine

An unconfined compression test machine in which *strain-controlled* tests can be performed is shown in Fig. 16.2. The machine essentially consists of a top and a bottom loading plate. The bottom of a proving ring is attached to the top loading plate. The top of the proving ring is attached to a crossbar which, in turn, is fixed to two metal posts. The bottom loading plate can be moved up or down.

Figure 16.2. Unconfined compression testing machine.

16.4 Procedure

1. Obtain a soil specimen for the test. If it is an undisturbed specimen, it has to be trimmed to the proper size by using the specimen trimmer. For classroom laboratory work, specimens at various moisture contents can be prepared using a Harvard miniature compaction device. The cylindrical soil specimen should have a height-to-diameter ratio L/D of between 2 and 3. In many instances, specimens with diameters of 1.4 in. (35.56 mm) and heights of 3.5 in. (88.9 mm) are used.

2. Measure the diameter D and length L of the specimen and determine the mass of the specimen.

3. Place the specimen centrally between the two loading plates of the unconfined compression testing machine. Move the top loading plate very carefully just to touch the top of the specimen. Set the proving ring dial gauge to zero. A dial gauge [each small division of the gauge should be equal to 0.001 in. (0.0254 mm) of vertical travel] should be attached to the unconfined compression testing machine to record the vertical upward movement (i.e., compression of specimen during testing) of the bottom loading plate. Set this dial gauge to zero.

Figure 16.3. Soil specimen after failure—by shearing.

4. Turn the machine on. Record loads (i.e., proving ring dial gauge readings) and the corresponding specimen deformations. During the load application, adjust the rate of *vertical strain* to 1/2–2% per minute. The rate of strain should be choosen such that the time to failure does not exceed about 15 minutes. At the initial stage of the test, readings are usually taken every 0.01 in. (0.254 mm) of specimen deformation. However, this can be varied to every 0.02 in. (0.508 mm) of specimen deformation at a later stage of the test, when the load–deformation curve begins to flatten out.

5. Continue taking readings until:
   a. Load reaches a peak and then decreases; or
   b. Load reaches a maximum value and remains approximately constant thereafter (take about 5 readings after it reaches its peak value); or
   c. Deformation of the specimen is past 15% strain before reaching the peak. This may happen in the case of soft clays.

   Figure 16.3 shows a soil specimen after failure by shearing and Fig. 16.4 shows the failure of the specimen by bulging.

6. Unload the specimen by lowering the bottom loading plate.

7. Remove the specimen from between the two loading plates.

8. Draw a free-hand sketch of the specimen after failure. Show the nature of the failure.

9. Put the specimen in a porcelain evaporating dish and determine the moisture content (after drying it in an oven to a constant weight).

Figure 16.4. Soil specimen after failure—by bulging.

16.5 Calculations

For each set of readings (refer to Table 16.1):

1. Calculate the vertical strain (column 2),

$$\varepsilon = \frac{\Delta L}{L} \tag{16.4}$$

> where ΔL = total vertical deformation of specimen
> L = original length of specimen

2. Calculate the vertical load on the specimen (column 4),

$$\text{load} = [\text{proving ring dial reading (column 3)}] \times (\text{calibration factor}) \tag{16.5}$$

3. Calculate the corrected area of the specimen (column 5),

$$A_c = \frac{A_0}{1 - \varepsilon} \tag{16.6}$$

where A_0 is the initial area of cross section of the specimen, $= (\pi/4)\,D^2$.

Table 16.1. Unconfined Compression Test

Description of soil _____Light brown clay_____ Specimen no. ___3___

Location _____Trinity Boulevard_____

Moist mass of specimen _____149.8_____ g Moisture content _____12_____ %

Length of specimen L _____3_____ in. Diameter of specimen _____1.43_____ in.

Proving ring calibration factor: 1 div. = ___0.264___ lb Area $A_0 = \frac{\pi}{4}D^2 =$ ___1.605___ in.^{2}

Tested by _____ Date _____

Specimen Deformation ΔL (in.) (1)	Vertical Strain $\varepsilon = \frac{\Delta L}{L}$ (2)	Proving Ring Dial Reading (no. of small div.) (3)	Load P (Column 3 × calibration factor) (lb) (4)	Corrected Area $A_c = \frac{A_0}{1-\varepsilon}$ (in.^{2}) (5)	Stress $\sigma = \frac{\text{Column 4}}{\text{Column 5}}$ (lb/in.^{2}) (6)
0	0	0	0	1.605	0
0.01	0.0033	12	3.168	1.611	1.966
0.02	0.0067	38	10.032	1.617	6.205
0.03	0.01	52	13.728	1.622	8.462
0.04	0.013	58	15.312	1.628	9.407
0.06	0.02	67	17.688	1.639	10.793
0.08	0.027	74	19.536	1.650	11.840
0.10	0.033	78	20.592	1.661	12.394
0.12	0.04	81	21.384	1.673	12.782
0.14	0.047	83	21.912	1.685	13.007
0.16	0.053	85	22.440	1.697	13.227
0.18	0.06	86	22.704	1.709	13.288
0.20	0.067	86	22.704	1.721	13.194
0.24	0.08	84	22.176	1.746	12.703
0.28	0.093	83	21.912	1.771	12.370
0.32	0.107	82	21.912	1.798	12.041
0.36	0.12	81	21.384	1.825	11.717

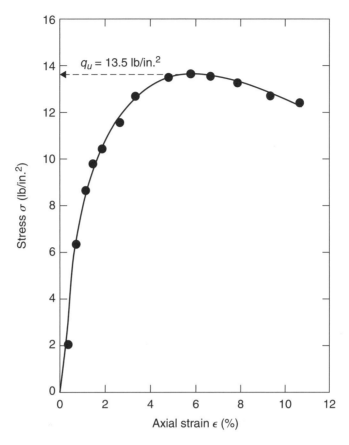

Figure 16.5. Plot of σ vs. ϵ (%) for the test results reported in Table 16.1.

4. Calculate the stress σ on the specimen (column 6),

$$\sigma = \frac{\text{load}}{A_c} = \frac{\text{column 4}}{\text{column 5}} \tag{16.7}$$

16.6 Graphs

Plot the graph of stress σ (column 6) versus axial strain ϵ, in percent (column 2 × 100). Determine the peak stress from this graph. This is the unconfined compression strength q_u of the specimen. (*Note:* If 15% strain occurs before the peak stress, then the stress corresponding to 15% strain should be taken as q_u.) A sample calculation and graph are shown in Table 16.1 and Fig. 16.5.

16.7 General Comments

1. In the determination of unconfined compression strength it is better to conduct tests on two to three identical specimens. The average value of q_u is the representative value.
2. Based on the value of q_u, the consistency of a cohesive soil is as shown in Table 16.2.

Table 16.2. Consistency of Cohesive Soils

Consistency	q_u (lb/ft^2)
Very soft	0–500
Soft	500–1000
Medium	1000–2000
Stiff	2000–4000
Very stiff	4000–8000

Table 16.3. Sensitivity of Clay

Sensitivity S_t	Description
1–2	Slightly sensitive
2–4	Medium sensitivity
4–8	Very sensitive
8–16	Slightly quick
16–32	Medium quick
32–64	Very quick
>64	Extra quick

3. For many naturally deposited clayey soils the unconfined compression strength is greatly reduced when the soil is tested after remolding without any change in moisture content. This is referred to as *sensitivity* and can be defined as

$$S_t = \frac{q_{u(\text{undisturbed})}}{q_{u(\text{remolded})}} \tag{16.8}$$

The sensitivity of most clays ranges from 1 to 8. Based on the magnitude of S_t, clays can be described as shown in Table 16.3.

17
Consolidation Test

17.1 Introduction

Consolidation is the process of time-dependent settlement of saturated clayey soil when subjected to increased loading. In this chapter the procedure of a one-dimensional laboratory consolidation test will be described, and the methods of calculation to obtain the void ratio versus pressure curve (e versus $\log p$), the preconsolidation pressure p_c, and the coefficient of consolidation c_v will be outlined.

17.2 Equipment

1. Consolidation test unit
2. Specimen trimming device
3. Wire saw
4. Balance sensitive to 0.01 g
5. Stopwatch
6. Moisture can
7. Oven

17.3 Consolidation Test Unit

The consolidation test unit consists of a consolidometer and a loading device. The consolidometer can be either floating ring or fixed ring (see Fig. 17.1). The floating-ring consolidometer usually consists of a brass ring in which the soil specimen is placed. One porous stone is placed at the top of the specimen and another porous stone at the bottom. The soil specimen in the ring with the two porous stones is placed on a base plate. A plastic ring surrounding the specimen fits into a groove on the base plate. Load is applied through a loading head, which is placed on the top porous stone. In the floating-ring consolidometer,

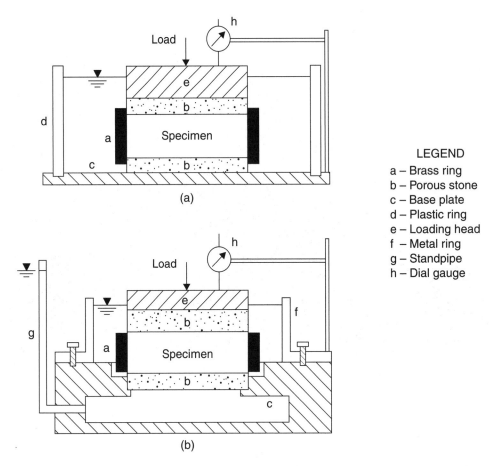

Figure 17.1. Schematic diagrams of consolidometer. (a) Floating-ring consolido-meter. (b) Fixed-ring consolidometer.

compression of the soil specimen occurs from the top and bottom toward the center. The fixed-ring consolidometer essentially consists of the same components, i.e., a hollow base plate, two porous stones, a brass ring to hold the soil specimen, and a metal ring that can be fixed tightly to the top of the base plate. The ring surrounds the soil specimen. A standpipe is attached to the side of the base plate. This can be used for determining the permeability of a soil. In the fixed-ring consolidometer the compression of the specimen occurs from the top toward the bottom.

The specifications for the loading devices of the consolidation test unit vary depending on the manufacturer. Figure 17.2 shows one type of loading device, and Fig. 17.3 is a closeup view of a consolidometer.

During the consolidation test, when load is applied to the soil specimen, the nature of the variations of the side friction between the surrounding brass ring and the specimen is different for the fixed-ring consolidometer and the floating-ring consolidometer, which is

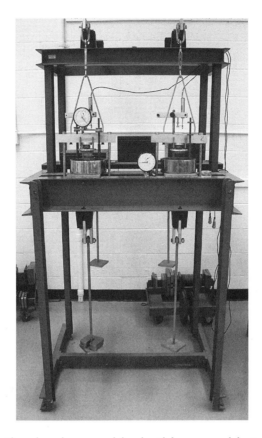

Figure 17.2. Consolidation load assembly. In this assembly two specimens can be tested simultaneously. Lever-arm ratio for loading is 1:10.

shown in Fig. 17.4. In most cases a side friction of 10% of the applied load is a reasonable estimate.

17.4 Procedure

1. Prepare a soil specimen for the test by trimming an undisturbed natural sample obtained in shelby tubes. The shelby tube sample should be about $1/4 - 1/2$ in. (6.35–12.7 mm) larger in diameter than the specimen diameter to be prepared for the test. (*Note:* For classroom instruction purposes, a specimen can be molded in the laboratory.)
2. Collect some excess soil that has been trimmed in a moisture can for determination of the moisture content.
3. Collect some of the excess soil trimmed in Step 1 for determination of the specific gravity G_s of the soil solids.
4. Determine the mass M_1 of the consolidation ring, in grams.

Figure 17.3. Close-up view of consolidometer during testing.

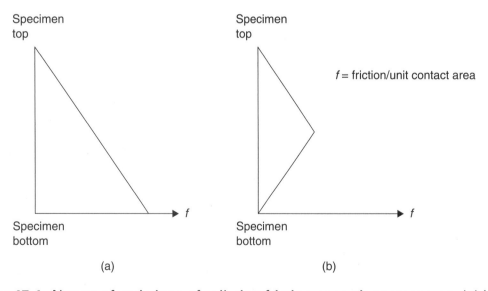

f = friction/unit contact area

(a)

(b)

Figure 17.4. Nature of variations of soil–ring friction per unit contact area. (a) In fixed-ring consolidometer: (b) In floating-ring consolidometer.

5. Place the soil specimen in the consolidation ring. Use the wire saw to trim the specimen flush with the top and bottom of the consolidation ring. Record the size of the specimen, i.e., height $H_{t(i)}$ and diameter D.

6. Determine the mass M_2 of the consolidation ring and the specimen, in grams.

7. Saturate the lower porous stone on the base of the consolidometer.

8. Place the soil specimen in the ring over the lower porous stone.

9. Place the upper porous stone on the specimen in the ring.

10. Attach the top ring to the base of the consolidometer.

11. Add water to the consolidometer to submerge the soil and keep it saturated. In the case of the fixed-ring consolidometer, the outside ring (which is attached to the top of the base) and the standpipe connection attached to the base should be kept filled with water. This needs to be done for the *entire period of the test.*

12. Place the consolidometer in the loading device.

13. Attach the vertical deflection dial gauge to measure the compression of the soil. It should be fixed in such as way that the dial is at the beginning of its release run. The dial gauge should be calibrated to read: 1 small div. = 0.0001 in. (0.00254 mm).

14. Apply load to the specimen such that the magnitude of pressure p on the specimen is $1/2$ ton/ft^2 (45.88 kN/m^2). Take the vertical deflection dial gauge readings at the following times t, counted from the time of load application: 0, 0.25, 1, 2.25, 4, 6.25, 9, 12.25, 20.25, 25, 36, 60, 120, 240, 480, and 1440 minutes (24 hours).

15. The next day add more load to the specimen such that the total pressure on the specimen reaches 1 ton/ft^2 (95.76 kN/m^2). Take the vertical deflection dial gauge reading at similar time intervals as stated in Step 14. (*Note:* Here we have $\Delta p/p = 1$, where Δp is the increase in pressure and p, the pressure before the increase.)

16. Repeat Step 15 for soil pressure magnitudes of 2 ton/ft^2 (191.52 kN/m^2), 4 ton/ft^2 (383.04 kN/m^2), and 8 ton/ft^2 (766.08 kN/m^2). (*Note:* $\Delta p/p = 1$.)

17. At the end of the test remove the soil specimen and determine its moisture content.

17.5 Calculations and Graphs

The calculation procedure for the test can be explained with reference to Tables 17.1 and 17.2 and Figs. 17.5, 17.6, and 17.7, which show the laboratory test results for a light brown clay.

1. Collect all the time versus vertical deflection dial readings. Table 17.1 shows the results of a pressure increase from $p = 2$ ton/ft^2 to $p + \Delta p = 4$ ton/ft^2.

2. Determine the time for 90% primary consolidation t_{90} from each set of time versus vertical deflection dial readings. An example is shown in Fig. 17.5, which is a plot of the results of vertical dial readings versus $\sqrt{\text{time}}$ given in Table 17.1. Draw a tangent AB to the initial consolidation curve. Measure the length BC. The abscissa of the point of intersection of line AD with the consolidation curve will give $\sqrt{t_{90}}$. Note that $CD = 1.15(BC)$. In Fig. 17.5, $\sqrt{t_{90}} = 4.75$ min.$^{0.5}$, so $t_{90} = (4.75)^2 = 22.56$ min. This technique is referred to as the *square-root-of-time curve fitting method* (Taylor, 1942).

Table 17.1. Time versus Vertical Dial Reading—Consolidation Test

Description of soil _____ *Light brown clay* _____

Location _____ *Summit Drive* _____

Tested by _____ Date _____

Pressure on specimen ____ 4 ____ ton/ft² Pressure on specimen _____ ton/ft²

Time after Load Application t (min)	$\sqrt{t}$ (min)$^{0.5}$	Vertical Dial Reading (in.)	Time after Load Application t (min)	$\sqrt{t}$ (min)$^{0.5}$	Vertical Dial Reading (in.)
0	0	0.0638			
0.25	0.5	0.0654			
1.0	1.0	0.0691			
2.25	1.5	0.0739			
4.0	2.0	0.0795			
6.25	2.5	0.0833			
9.0	3.0	0.0868			
12.25	3.5	0.0898			
16.0	4.0	0.0922			
20.25	4.5	0.0941			
25	5.0	0.0954			
36	6.0	0.0979			
60	7.75	0.1004			
120	10.95	0.1019			
240	15.49	0.1029			
480	21.91	0.1048			
1440	37.95	0.1059			

Table 17.2. Pressure, Void Ratio, and Calculation of Coefficient of Consolidation—Consolidation Test

Description of soil ___Light brown clay___ Location ___

Specimen diameter __2.5__ in. Initial specimen height $H_{t(i)}$ __1__ in. Height of solids H_s __0.539__ in.

Moisture content: Beginning of test __30.8__ % End of test __32.1__ % Mass of dry soil specimen __116.74__ g G_s __2.72__

Tested by ___ Date ___

Pressure p (ton/ft²) (1)	Final Dial Reading (in.) (2)	Change in Specimen Height ΔH (in.) (3)	Final Specimen Height $H_{t(f)}$ (in.) (4)	Height of Voids H_v (in.) (5)	Final void ratio e (6)	Average Height during Consolidation $H_{t(av)}$ (in.) (7)	Fitting Time (s) t_{90} (8)	Fitting Time (s) t_{50} (9)	$c_v \times 10^3$ (in.²/sec) from t_{90} (10)	$c_v \times 10^3$ (in.²/sec) from t_{50} (11)
0	0.200		1.000	0.4610	0.855					
		0.0083				0.9959	302	68.7	0.696	0.711
1/2	0.0283		0.9917	0.4527	0.840					

continued

133

Table 17.2. Continued

Pressure p (ton/ft²) (1)	Final Dial Reading (in.) (2)	Change in Specimen Height ΔH (in.) (3)	Final Specimen Height $H_{t(f)}$ (in.) (4)	Height of Voids H_v (in.) (5)	Final void ratio e (6)	Average Height during Consolidation $H_{t(av)}$ (in.) (7)	Fitting Time (s) t_{90} (8)	Fitting Time (s) t_{50} (9)	$c_v \times 10^3$ (in.²/sec) from t_{90} (10)	$c_v \times 10^3$ (in.²/sec) from t_{50} (11)
		0.0073				0.9881	308	56.0	0.672	0.859
1	0.0356		0.9844	0.4454	0.826					
		0.0282				0.9703	492	144	0.406	0.322
2	0.0638		0.9562	0.4172	0.774					
		0.0421				0.9352	1354	240	0.137	0.179
4	0.1059		0.9141	0.3751	0.696					
		0.0455				0.8914	1102	249	0.153	0.133
8	0.1514		0.8686	0.3296	0.612					

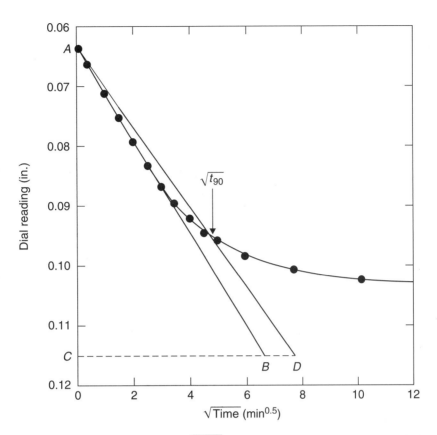

Figure 17.5. Plot of dial reading vs. √time for test results reported in Table 17.1. Determination of t_{90} by square-root-of-time curve fitting method. Note: $CD = 1.15(BC)$.

3. Determine the time for 50% primary consolidation t_{50} from each set of time versus vertical deflection dial readings. The procedure for this is shown in Fig. 17.6, which is a semilog plot (vertical dial readings on a natural scale and time on a log scale) for the set of readings shown in Table 17.1. Project the straight-line portion of the primary consolidation downward and the straight-line portion of the secondary consolidation backward. The point of intersection of these two lines is A. The vertical dial reading corresponding to A is d_{100} (dial reading at 100% primary consolidation). Select times t_1 and $t_2 = 4t_1$. (*Note:* t_1 and t_2 should be within the top curved portion of the consolidation plot.) Determine the difference X in the dial readings between times t_1 and t_2. Plot line BC, which is vertically a distance X above the point on the consolidation curve corresponding to time t_1. The vertical dial gauge corresponding to line BC is d_0, i.e., the reading for 0% consolidation. Determine the dial gauge reading corresponding to 50% primary consolidation,

$$d_{50} = \frac{d_0 + d_{100}}{2} \tag{17.1}$$

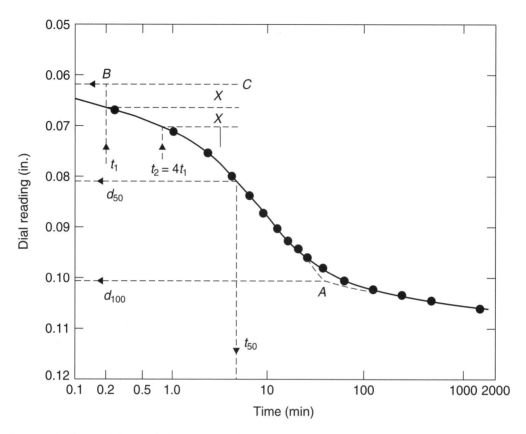

Figure 17.6. Logarithm-of-time curve fitting method for laboratory results reported in Table 17.1.

The time corresponding to d_{50} on the consolidation curve is t_{50}. This is the *logarithm-of-time curve fitting method* (Casagrande and Fadum, 1940). In Figure 17.6, $t_{50} = 4.0$ min.

4. Complete the experimental data in columns 1, 2, 8, and 9 of Table 17.2. Columns 1 and 2 are obtained from time–dial reading tables (such as Table 17.1), and columns 8 and 9 are obtained from Steps 2 and 3, respectively.

5. Determine the height H_s of the solids of the specimen in the mold (see top of Table 17.2),

$$H_s = \frac{M_s}{\left(\frac{\pi}{4}D^2\right)G_s\rho_w} \tag{17.2}$$

where M_s = dry mass of soil specimen
D = diameter of specimen
G_s = specific gravity of soil solids
ρ_w = density of water

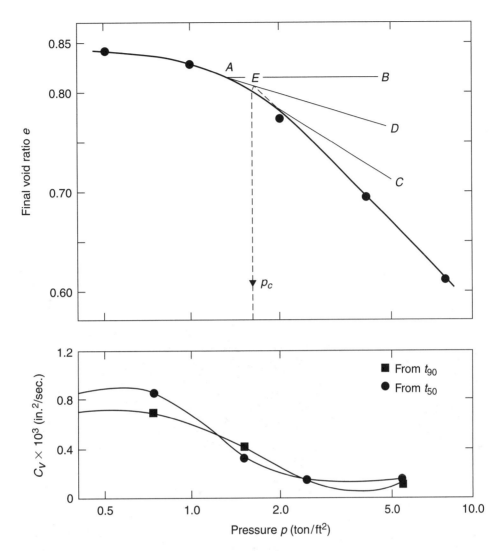

Figure 17.7. Plots of void ratio and coefficient of consolidation use pressure for soil reported in Table 17.2.

6. In Table 17.2 determine the change in height ΔH of the specimen due to load increments from p to $p + \Delta p$ (column 3). For example,

$$p = 1/2 \text{ ton/ft}^2, \quad \text{final dial reading} = 0.0283 \text{ in.}$$

$$p + \Delta p = 1 \text{ ton/ft}^2, \quad \text{final dial reading} = 0.0356 \text{ in.}$$

Thus,

$$\Delta H = 0.0356 - 0.0283 = 0.0073 \text{ in.}$$

7. Determine the final specimen height $H_{t(f)}$ at the end of consolidation due to a given load (column 4, Table 17.2). For example, in Table 17.2 $H_{t(f)}$ at $p = 1/2 \text{ ton/ft}^2$ is 0.9917.

ΔH from $p = 1/2$ ton/ft^2 to 1 ton/ft^2 is 0.0073. So $H_{t(f)}$ at $p = 1$ ton/ft^2 equals $0.9917 - 0.0073 = 0.9844$ in.

8. Determine the height H_v of the voids in the specimen at the end of consolidation due to a given loading p (column 5, Table 17.2),

$$H_v = H_{t(f)} - H_s \tag{17.3}$$

9. Determine the final void ratio at the end of consolidation for each loading p (column 6, Table 17.2),

$$e = \frac{H_v}{H_s} = \frac{\text{column 5}}{H_s} \tag{17.4}$$

10. Determine the average specimen height $H_{t(av)}$ during consolidation for each incremental loading (column 7, Table 17.2). For example, in Table 17.2 the value of $H_{t(av)}$ between $p = 1/2$ ton/ft^2 and $p = 1$ ton/ft^2 is

$$\frac{H_{t(f)} \text{ at } p = \frac{1}{2} \text{ ton/ft}^2 + H_{t(f)} \text{ at } p = 1 \text{ ton/ft}^2}{2} = \frac{0.9917 + 0.9844}{2} = 0.9881 \text{ in.}$$

11. Calculate the coefficient of consolidation c_v (column 10, Table 17.2) from t_{90} (column 8),

$$T_v = \frac{c_v t}{H^2} \tag{17.5}$$

where T_v = time factor, $t_{90} = 0.848$
H = maximum length of drainage path, $= H_{t(av)}/2$, (since specimen is drained at top and bottom)

Thus,

$$c_v = \frac{0.848 H_{t(av)}^2}{4 t_{90}} \tag{17.6}$$

12. Calculate the coefficient of consolidation c_v (column 11, Table 17.2) from t_{50} (column 9),

$$T_{v(50\%)} = 0.197 = \frac{c_v t_{50}}{H^2} = \frac{c_v t_{50}}{(H_{t(av)}/2)^2}$$

Then

$$c_v = \frac{0.197 (H_{t(av)})^2}{4 t_{50}} \tag{17.7}$$

For example, from $p = 1/2$ ton/ft^2 to $p = 1$ ton/ft^2, $H_{t(av)} = 0.9881$ in.; $t_{50} = 56.0$ s. So

$$c_v = \frac{0.197 (0.9881)^2}{4 \times 56} = 0.859 \times 10^{-3} \text{ in.}^2/\text{s}$$

13. Plot a semilogarithmic graph of pressure versus final void ratio (column 1 versus column 6, Table 17.2). Pressure p is plotted on the log scale and the final void ratio on the linear scale. As an example, the results of Table 17.2 are plotted in Fig. 17.7. (*Note:* The plot of e versus log p has a curved upper portion and then a linear relationship.)
14. Calculate the compression index C_c. This is the slope of the linear portion of the e versus log p plot (Step 13). In Fig. 17.7 (also see Table 17.2),

$$C_c = \frac{e_1 - e_2}{\log(p_2/p_1)} = \frac{0.696 - 0.612}{\log(8/4)} = 0.279$$

Table 17.3. Empirical Relations for C_c

Reference	Relation	Comments
Terzaghi and Peck (1967)	$C_c = 0.009(LL - 10)$ $C_c = 0.007(LL - 10)$ $LL = $ liquid limit (%)	Undisturbed clay Remolded clay
Azzouz et al. (1976)	$C_c = 0.01w_N$ $w_N = $ natural moisture content (%) $C_c = 0.0046(LL - 9)$ $LL = $ liquid limit (%) $C_c = 1.21 + 1.005(e_0 - 1.87)$ $e_0 = $ in situ void ratio $C_c = 0.208e_0 + 0.0083$ $e_0 = $ in situ void ratio $C_c = 0.0115w_N$ $w_N = $ natural moisture content (%)	Chicago clay Brazilian clay Motley clays from Sao Paulo City Chicago clay Organic soil, peat
Nacci et al. (1975)	$C_c = 0.02 + 0.014(PI)$ $PI = $ plasticity index (%)	North Atlantic clay
Rendon–Herrero (1980)	$C_c = 0.141G_s^{1.2}\left(\dfrac{1 + e_0}{G_s}\right)^{2.38}$ $G_s = $ specific gravity of soil solids $e_0 = $ in situ void ratio	
Nagaraj and Murty (1985)	$C_c = 0.2343\left(\dfrac{LL}{100}\right)G_s$ $G_s = $ specific gravity of soil solids $LL = $ liquid limit (%)	

15. On the semilogarithmic graph (Step 13), using the same horizontal scale (the scale for p), plot the values of c_v (columns 10 and 11, Table 17.2). As an example, the values determined in Table 17.2 are plotted in Fig. 17.7.) (*Note*: c_v is plotted on the linear scale corresponding to the average value of p, i.e., $(p_1 + p_2)/2$.)

16. Determine the *preconsolidation pressure* p_c. The procedure can be explained with the aid of the e versus log p graph shown in Fig. 17.7 (Casagrande, 1936). First determine point A, which is the point on the e versus log p plot that has the smallest radius of curvature. Draw a horizontal line AB. Draw a line AD, which is the *bisector* of angle BAC. Project the straight-line portion of the e versus log p plot backward to meet line AD at E. The pressure corresponding to point E is the preconsolidation pressure. In Fig. 17.7, $p_c = 1.6$ ton/ft^2.

17.6 General Comments

The magnitude of the compression index C_c varies from soil to soil. Many correlations for C_c have been proposed in the past for various types of soils. A summary of these correlations is given in Table 17.3.

18
Triaxial Tests in Clay

18.1 Introduction

In Chapters 15 and 16 some aspects of the shear strength of soils were discussed. The relationship for shear strength s of a soil was given in Chapter 16 as

$$s = c' + \sigma' \tan \phi' \qquad (18.1)$$

where $c' =$ cohesion
$\phi' =$ drained angle of friction
$\sigma' =$ effective normal stress

Also, for undrained condition (that is, $\phi = 0$ condition)

$$s = c_u \qquad (18.2)$$

where $c_u =$ undrained cohesion

The triaxial compression test is a more sophisticated test procedure for determining the shear strength of a soil. In general, with triaxial equipment three types of common tests can be conducted, which are listed in Table 18.1. Both the unconsolidated-undrained test and the consolidated-undrained test will be described in this chapter.

18.2 Equipment

1. Triaxial cell
2. Strain-controlled compression machine
3. Specimen trimmer
4. Wire saw
5. Vacuum source
6. Oven

Table 18.1. Triaxial Compression Tests

Test Type	Parameters Determined
Unconsolidated-undrained (U-U)	$c_u \ (\phi = 0)$
Consolidated-drained (C-D)	c', ϕ'
Consolidated-undrained (C-U)	$c', \phi', \bar{A}$

$\bar{A}$ = pore water pressure parameter.

7. Calipers
8. Evaporating dish
9. Rubber membrane
10. Membrane stretcher

18.3 Triaxial Cell and Loading Arrangement

Figure 18.1 shows the schematic diagram of a triaxial cell. It consists mainly of a bottom base plate, a Lucite cylinder, and a top cover plate. A bottom platen is attached to the base plate. A porous stone is placed over the bottom platen, over which the soil specimen is placed. A porous stone and a platen are placed on top of the specimen. The specimen is enclosed inside a thin rubber membrane. Inlet and outlet tubes for specimen saturation and drainage are provided through the base plate. Appropriate valves are attached to these tubes to shut off the openings when desired. There is an opening in the base plate through which water (or glycerine) can be allowed to flow to fill the cylindrical chamber. A hydrostatic chamber pressure σ_3 can be applied to the specimen through the chamber fluid. An added axial stress $\Delta\sigma$, applied to the top of the specimen, can be provided using a piston.

During the test, the triaxial cell is placed on the platform of a strain-controlled compression machine. The top of the piston of the triaxial chamber is attached to a proving ring. The proving ring is attached to a crossbar that is fixed to two metal posts. The platform of the compression machine can be raised (or lowered) at desired rates, thereby raising (or lowering) the triaxial cell. During compression, the load on the specimen can be obtained from the proving ring readings and the corresponding specimen deformation from a dial gauge [1 small div. = 0.001 in. (0.0254 mm)].

The connections to the soil specimen can be attached to a burette or a pore-water pressure measuring device to measure, respectively, the volume change of the specimen or the excess pore-water pressure during the test.

Triaxial equipment is costly, depending on the accessories attached to it. For that reason only general test procedures will be outlined here. For locating the various components of an assembly, students will need the help of their instructors.

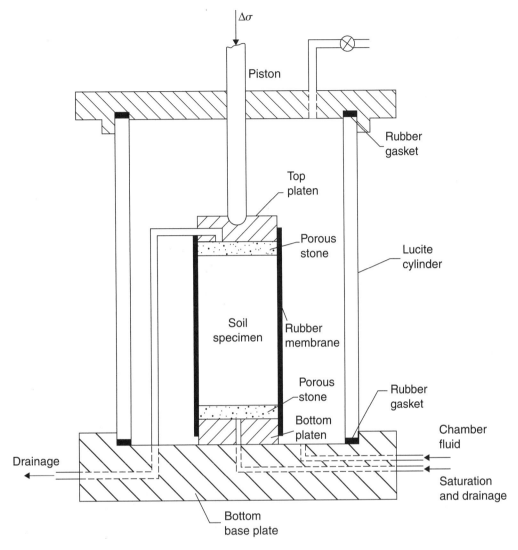

Figure 18.1. Schematic diagram of triaxial cell.

18.4 Triaxial Specimen

Triaxial specimens most commonly used are about 2.8 in. in diameter × 6.5 in. in length (71.1-mm diameter × 165.1-mm length) or 1.4 in. in diameter × 3.5 in. in length (35.6-mm diameter × 88.9-mm length). In any case, the length-to-diameter ratio L/D should be between 2 and 2.5. For tests on undisturbed natural soil samples collected in Shelby tubes, a specimen trimmer may need to be used to prepare a specimen of desired dimensions. Depending on the triaxial cell at hand, for classroom use, remolded specimens can be prepared with Harvard miniature compaction equipment.

After the specimen is prepared, obtain its length L_0 and diameter D_0. The length should be measured four times about 90 degrees apart. The average of these four values should be equal to L_0. To obtain the diameter, take four measurements at the top, four at the middle, and four at the bottom of the specimen. The average of these twelve measurements is D_0.

18.5 Placement of Specimen in Triaxial Cell

1. Boil the two porous stones to be used with the specimen.
2. De-air the lines connecting the base of the triaxial cell.
3. Attach the bottom platen to the base of the cell.
4. Place the bottom porous stone (moist) over the bottom platen.
5. Take a thin rubber membrane of appropriate size to fit the specimen snugly. Take a membrane stretcher, which is a brass tube having an inside diameter of about 1/4 in. (≈ 6 mm) larger than the specimen diameter (Fig. 18.2). The membrane stretcher can be connected to a vacuum source. Fit the membrane to the inside of the membrane stretcher and lap the ends of the membrane over the stretcher. Then apply the vacuum. This will make the membrane form a smooth cover inside the stretcher.
6. Slip the soil specimen inside the stretcher with the membrane (Step 5). The inside of the membrane may be moistened for ease in slipping in the specimen. Now release the vacuum and unroll the membrane from the ends of the stretcher.
7. Place the specimen (Step 6) on the bottom porous stone (which is placed on the bottom platen of the triaxial cell) and stretch the bottom end of the membrane around the porous stone and bottom platen. At this time place the top porous stone (moist) and the top platen on the specimen and stretch the top of the membrane over it. For airtight

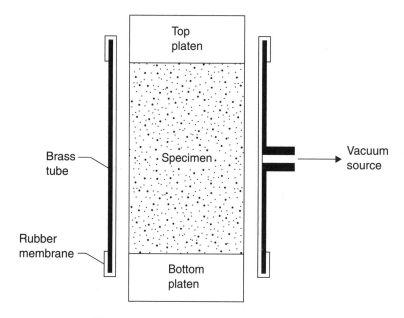

Figure 18.2. Membrane stretcher.

seals it is always a good idea to apply some silicone grease around the top and bottom platens before the membrane is stretched over them.

8. Using some rubber bands, fasten the membrane tightly around the top and bottom platens.

9. Connect the drainage line leading from the top platen to the base of the triaxial cell.

10. Place the Lucite cylinder and the top of the triaxial cell on the base plate to complete the assembly.

Notes:

- In the triaxial cell the specimen can be saturated by connecting the drainage line leading to the bottom of the specimen to a saturation reservoir. During this process the drainage line leading from the top of the specimen is kept open to the atmosphere. The saturation of clay specimens takes a fairly long time.

- For the unconsolidated-undrained test, if specimen saturation is not required, non-porous plates can be used instead of porous stones at the top and bottom of the specimen.

18.6 Unconsolidated-Undrained Test

Procedure

1. Place the triaxial cell (with the specimen inside) on the platform of the compression machine.

2. Make proper adjustments so that the piston of the triaxial cell just rests on the top platen of the specimen

3. Fill the chamber of the triaxial cell with water. Apply a hydrostatic pressure σ_3 to the specimen through the chamber fluid. Wait for about 10 min after the application of chamber-confining pressure to allow the specimen to stabilize. (*Note:* All drainage to and from the specimen should be closed *now* so that drainage from the specimen does not occur.)

4. Check for proper contact between the piston and the top platen on the specimen. Zero the dial gauge of the proving ring and the gauge used for measuring the vertical compression of the specimen. Set the compression machine for a strain rate of about 0.5%/min and turn on the switch.

5. Take initial proving ring dial readings for vertical compression intervals of 0.01 in. (0.254 mm). This interval can be increased to 0.02 in. (0.508 mm) or more later on, when the rate of increase of the load on the specimen decreases. The proving ring readings will increase to a peak value and then may decrease or remain approximately constant. Take about four to five readings after the peak point.

6. After completion of the test, reverse the compression machine, lower the triaxial cell, and then turn off the machine. Release the chamber pressure and drain the water in the

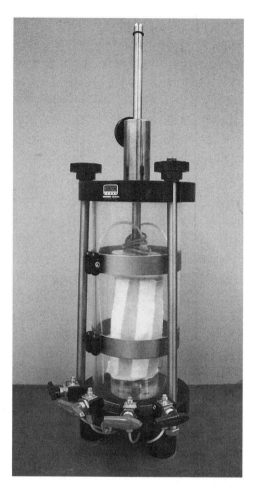

Figure 18.3. Specimen inside a triaxial chamber after completion of test.

triaxial cell. Figure 18.3 shows a specimen inside a triaxial chamber after completion of the test. Then remove the specimen and determine its moisture content.

Calculations

The calculation procedure can be explained by referring to Tables 18.2 and 18.3, which present the results of an unconsolidated-undrained triaxial test on a dark brown silty clay specimen. Referring to Table 18.2,

1. Calculate the final moisture content w of the specimen (line 3),

$$w\,(\%) = \frac{\text{moist mass of specimen } M_1 - \text{ dry mass of specimen } M_2}{\text{dry mass of specimen } M_2} \times 100$$

$$= \frac{\text{line 1} - \text{line 2}}{\text{line 2}} \times 100 \tag{18.3}$$

Table 18.2. Unconsolidated-Undrained Triaxial Test—Preliminary Data

Description of soil _____ *Dark brown silty clay* _____   Specimen no. ___ *8* ___

Location _____

Tested by _____   Date _____

Item	Quantity
1. Moist mass of specimen (end of test), M_1	*185.68 g*
2. Dry mass of specimen, M_2	*151.80 g*
3. Moisture content (end of test), $w\ (\%) = \dfrac{M_1 - M_2}{M_2} \times 100$	*22.3%*
4. Initial average length of specimen, L_0	*3.52 in.*
5. Initial average diameter of specimen, D_0	*1.41 in*
6. Initial area, $A_0 = \dfrac{\pi}{4}D_0^2$	*1.56 in.^{2}*
7. Specific gravity of soil solids, G_s	*2.73*
8. Final degree of saturation	*98.2%*
9. Cell confining pressure, σ_3	*15 lb/in.^{2}*
10. Proving ring calibration factor	*0.37 lb.div.*

2. Calculate the initial area of the specimen (line 6),

$$A_0 = \frac{\pi}{4}D_0^2 = \frac{\pi}{4}(\text{line 5})^2 \tag{18.4}$$

3. Now refer to Table 18.3 and calculate the vertical strain (column 2),

$$\varepsilon = \frac{\Delta L}{L_0} = \frac{\text{column 1}}{\text{line 4 (Table 18.1)}} \tag{18.5}$$

where ΔL is the total deformation of the specimen at any time.

4. Calculate the piston load on the specimen (column 4),

$$P = \text{proving ring dial reading} \ \times \ \text{calibration factor}$$
$$= \text{column 3} \times \text{line 10 (Table 18.1)} \tag{18.6}$$

5. Calculate the corrected area A of the specimen (column 5),

$$A = \frac{A_0}{1 - \varepsilon} = \frac{\text{line 6 (Table 18.1)}}{1 - \text{column 2}} \tag{18.7}$$

6. Calculate the deviatory stress (or piston stress) $\Delta\sigma$ (column 6),

$$\Delta\sigma = \frac{P}{A} = \frac{\text{column 4}}{\text{column 5}} \tag{18.8}$$

Graphs

1. Draw a graph of the axial strain versus deviatory stress $\Delta\sigma$. As an example, the results of Table 18.3 are plotted in Fig. 18.4. From this graph obtain the value of $\Delta\sigma$ at failure ($\Delta\sigma = \Delta\sigma_f$).

2. The minor principal stress (*total*) on the specimen at failure is σ_3 (that is, the chamber confining pressure). Calculate the major principal stress (*total*) at failure,

$$\sigma_1 = \sigma_3 + \Delta\sigma_f \tag{18.9}$$

3. Draw Mohr's circle with σ_1 and σ_3 as the major and minor principal stresses. The radius of Mohr's circle is equal to c_u. The results of the test reported in Table 18.3 and Fig. 18.4 are plotted in Fig. 18.5.

General Comments

1. For saturated clayey soils, the unconfined compression test discussed in Chapter 16 is a special case of the U-U test. For the unconfined compression test, $\sigma_3 = 0$. However, the quality of the results obtained from U-U tests is superior.

2. Figure 18.6 shows the nature of Mohr's envelope obtained from U-U tests with varying degrees of saturation. For saturated specimens the value of $\Delta\sigma_f$, and thus c_u, is constant, irrespective of the chamber confining pressure σ_3. So Mohr's envelope is a horizontal line ($\phi = 0$). For soil specimens with degrees of saturation less than 100%, Mohr's envelope is curved and falls above the $\phi = 0$ line.

3. In Section 18.6. (Procedure), Steps 4 and 5, the general strain rate and the test duration were outlined. However, ASTM test designation D-2850 is more specific about it. A summary of the guidelines of D-2850 follows:

   - The strain rate of plastic material should be about 1%/min. However, for brittle material it should be about 0.3%/min.
   - The axial load (i.e., the deviator stress) should continue to be increased until either at least 15% of the axial strain is reached, or the deviator stress has peaked and dropped 20% or the axial strain has reached 5% beyond the strain at which the peak deviator stress occurred.

Table 18.3. Axial Stress–Strain Calculation—Unconsolidated-Undrained Triaxial Test

Specimen Deformation ΔL (in.) (1)	Vertical Strain $\varepsilon = \dfrac{\Delta L}{L_0}$ (2)	Proving Ring Dial Reading (no. of small div.) (3)	Piston Load P (column 3 × calibration factor) (lb) (4)	Corrected Area $A = \dfrac{A_0}{1-\varepsilon}$ (in.^{2}) (5)	Deviatory Stress $\Delta\sigma = \dfrac{P}{A}$ (lb/in.^{2}) (6)
0	0	0	0	1.560	0
0.01	0.0028	3.5	1.295	1.564	0.828
0.02	0.0057	7.5	2.775	1.569	1.769
0.03	0.0085	11	4.07	1.573	2.587
0.04	0.0114	14	5.18	1.578	3.28
0.05	0.0142	18	6.66	1.582	4.210
0.06	0.0171	21	7.77	1.587	4.896
0.10	0.0284	31	11.47	1.606	7.142
0.14	0.0398	38	14.06	1.625	8.652
0.18	0.0511	44	16.28	1.644	9.903
0.22	0.0625	48	17.76	1.664	10.673
0.26	0.0739	52	19.24	1.684	11.425
0.30	0.0852	53	19.61	1.705	11.501
0.35	0.0994	52	19.24	1.735	11.109
0.40	0.1136	50	18.5	1.760	10.511
0.45	0.1278	49	18.13	1.789	10.134
0.50	0.1420	49	18.13	1.818	9.970

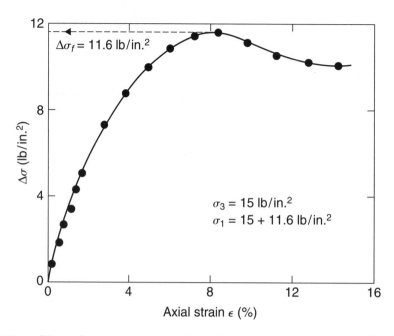

Figure 18.4. Plot of $\Delta\sigma$ against axial strain for test reported in Table 18.3.

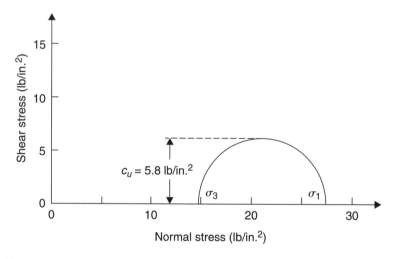

Figure 18.5. Total-stress Mohr's circle at failure for test reported in Table 18.3 and Fig. 18.4.

18.7 Consolidated-Undrained Test

Procedure

1. Place the triaxial cell with the saturated specimen on the compression machine platform and make adjustments so that the piston of the cell makes contact with the top platen of the specimen.

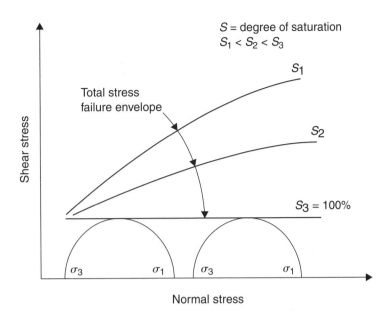

Figure 18.6. Nature of variations of total-stress failure envelopes with degree of saturation of soil specimen (undrained test).

2. Fill the chamber of the triaxial cell with water, and apply the hydrostatic pressure σ_3 to the specimen through the fluid.
3. The application of the chamber pressure σ_3 will cause an increase in the pore-water pressure in the specimen. For consolidation connect the drainage lines from the specimen to a calibrated burette and leave the lines open. When the water level in the burette becomes constant, it will indicate that the consolidation is complete. For a saturated specimen the volume change due to consolidation is equal to the volume of water drained from the burette. Record the volume of the drainage ΔV.
4. Now connect the drainage lines to the pore-pressure measuring device.
5. Check the contact between piston and top platen. Zero the proving ring dial gauge and the dial gauge that measures the axial deformation of the specimen.
6. Set the compression machine for a strain rate of about 0.5%/min and turn on the switch. When the axial load on the specimen is increased, the pore-water pressure in the specimen will also increase. Record the proving ring dial gauge reading and the corresponding excess pore-water pressure Δu in the specimen for every 0.01 in. (0.254 mm) or less of axial deformation. The proving ring dial gauge reading will increase to a maximum and then decrease or remain approximately constant. Take at least four to five readings after the proving ring dial gauge reaches the maximum value. According to ASTM test designation D-4767, the axial load (i.e., the deviator stress) should continue to be increased until either at least 15% of the axial strain is reached, or the deviator stress has peaked and dropped 20% or the axial strain has reached 5% beyond the strain at which the peak deviator stress occurred.

7. At the completion of the test, reverse the compression machine and lower the triaxial cell. Shut off the machine. Release the chamber pressure σ_3 and drain the water out of the triaxial cell.
8. Remove the tested specimen from the cell and determine its moisture content.
9. Repeat the test on one or two more similar specimens. Each specimen should be tested at a different value of σ_3.

Calculations and Graphs

The procedure for making the required calculations and plotting the graphs can be explained by referring to Tables 18.4 and 18.5 and Figs. 18.7 and 18.8.

First refer to Table 18.4.

1. Calculate the initial area of the specimen (line 5),

$$A_0 = \frac{\pi}{4}D_0^2 = \frac{\pi}{4}(\text{line 4})^2 \tag{18.10}$$

2. Calculate the initial volume of the specimen (line 6),

$$V_0 = A_0 L_0 = \text{line 5} \times \text{line 3} \tag{18.11}$$

3. Calculate the volume of the specimen after consolidation (line 9),

$$V_c = V_0 - \Delta V = \text{line 6} - \text{line 8} \tag{18.12}$$

where V_c is the final volume of the specimen.
4. Calculate the length L_c (line 10) and cross-sectional area A_c (line 11) of the specimen after consolidation,

$$L_c = L_0 \left(\frac{V_c}{V_0}\right)^{1/3} = \text{line 3} \times \left(\frac{\text{line 9}}{\text{line 6}}\right)^{1/3} \tag{18.13}$$

and

$$A_c = A_0 \left(\frac{V_c}{V_0}\right)^{2/3} = \text{line 5} \times \left(\frac{\text{line 9}}{\text{line 6}}\right)^{2/3} \tag{18.14}$$

Now refer to Table 18.5.
5. Calculate the axial strain (column 2),

$$\varepsilon = \frac{\Delta L}{L_c} = \frac{\text{column 1}}{\text{line 10 (Table 18.4)}} \tag{18.15}$$

where ΔL is the axial deformation.
6. Calculate the piston load P (column 4),

$$P = \text{proving ring dial reading (column 3)} \times \text{calibration factor} \tag{18.16}$$

Table 18.4. Preliminary Data—Consolidated-Undrained Triaxial Test

Description of soil _____ *Remolded grundite* _____ Sample no. ___2___

Location _____

Tested by _____ Date _____

Beginning of Test	
1. Moist unit weight of specimen (beginning of test)	$18.4 \ kN/m^3$
2. Moisture content (beginning of test)	35.35%
3. Initial length of specimen, L_0	$7.62 \ cm$
4. Initial diameter of specimen, D_0	$3.57 \ cm$
5. Initial area of specimen, $A_0 = \frac{\pi}{4}D_0^2$	$10.0 \ cm^2$
6. Initial volume of specimen, $V_0 = A_0L_0$	$76.2 \ cm^3$
After Consolidation of Saturated Specimen	
7. Cell consolidation pressure, σ_3	$392 \ kN/m^2$
8. Net drainage from specimen during consolidation, ΔV	$11.6 \ cm^3$
9. Volume of specimen after consolidation, $V_0 - \Delta V = V_c$	$76.2 - 11.6 = 64.6 \ cm^3$
10. Length of specimen after consolidation, $L_c = L_0 \left(\dfrac{V_c}{V_0}\right)^{2/3}$	$7.62 \left(\dfrac{64.6}{76.2}\right)^{1/3} = 7.212 \ cm$
11. Area of specimen after consolidation, $A_c = A_0 \left(\dfrac{V_c}{V_0}\right)^{2/3}$	$10 \left(\dfrac{64.6}{76.2}\right)^{2/3} = 8.96 \ cm^3$

7. Calculate the corrected area A (column 5),

$$A = \frac{A_c}{1 - \varepsilon} = \frac{\text{line 11 (Table 18.4)}}{1 - \text{column 2}} \tag{18.17}$$

Table 18.5. Axial Stress–Strain Calculation*—Consolidated-Undrained Triaxial Test

Proving ring calibration factor _____ *1.0713 N/dvi.* _____

Specimen Deformation ΔL (cm) (1)	Vertical Strain $\varepsilon = \frac{\Delta L}{L_c}$ (2)	Proving Ring Dial Reading (no. of small div.) (3)	Piston Load P (N) (4)	Corrected Area $A = \frac{A_c}{1-\varepsilon}$ (cm^2) (5)	Deviatory Stress $\Delta\sigma = \frac{P}{A}$ (kN/m^2) (6)	Excess Pore-Water Pressure Δu (kN/m^2) (7)	$\bar{A} = \frac{\Delta u}{\Delta \sigma}$ (8)
0	0	0	0	8.96	0	0	0
0.015	0.0021	15	16.07	8.98	17.90	2.94	0.164
0.038	0.0053	109	116.77	9.01	129.60	49.50	0.378
0.061	0.0085	135	144.63	9.04	159.99	74.56	0.466
0.076	0.0105	147	157.48	9.06	173.82	89.27	0.514
0.114	0.0158	172	184.26	9.11	202.26	111.83	0.553
0.152	0.0211	192	205.69	9.15	224.80	135.38	0.602
0.183	0.0254	205	219.62	9.19	238.98	148.13	0.620
0.229	0.0318	225	241.04	9.25	260.58	160.88	0.618
0.274	0.0380	236	252.83	9.31	271.57	167.75	0.618
0.315	0.0437	247	264.61	9.37	282.40	175.60	0.622
0.427	0.0592	265	283.89	9.52	298.20	176.58	0.592
0.457	0.0634	270	289.25	9.57	302.25	176.58	0.584
0.503	0.0697	278	297.82	9.63	309.26	176.58	0.570
0.549	0.0761	284	304.25	9.70	313.66	176.58	0.563
0.594	0.0824	287	307.46	9.76	315.02	176.58	0.561
0.653	0.0905	288	308.53	9.85	313.23	176.57	0.564
0.726	0.1007	286	306.39	9.96	307.62	160.88	0.523
0.853	0.183	275	294.61	10.16	289.97	163.82	0.565

* The results have been edited.

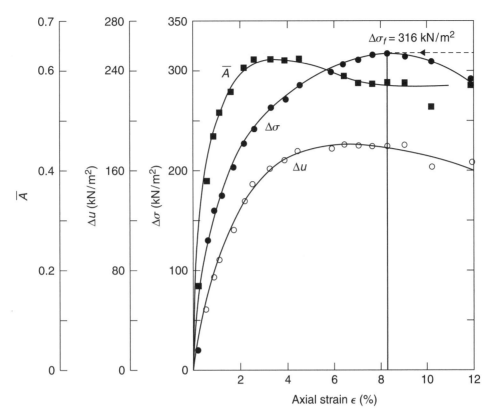

Figure 18.7. Plots of $\Delta\sigma$, Δu, and $\bar{A}$ use axial strain for consolidated drained test reported in Table 18.5

8. Determine the deviatory stress $\Delta\sigma$ (column 6),

$$\Delta\sigma = \frac{P}{A} = \frac{\text{column 4}}{\text{column 5}} \tag{18.18}$$

9. Determine the pore-water pressure parameter $\bar{A}$ (column 8),

$$\bar{A} = \frac{\Delta u}{\Delta\sigma} = \frac{\text{column 7}}{\text{column 6}} \tag{18.19}$$

10. Plot graphs for
    a. $\Delta\sigma$ versus $\in$
    b. Δu versus $\in$
    c. $\bar{A}$ versus $\in$
    As an example, the results of the calculations shown in Table 18.5 are plotted in Fig. 18.7.
11. From the $\Delta\sigma$ versus $\in$ graph, determine the maximum value of $\Delta\sigma = \Delta\sigma_f$ and the corresponding values of $\Delta u = \Delta u_f$ and $\bar{A} = \bar{A}_f$. In Fig. 18.7, $\Delta\sigma_f = 316$ kN/m^2 at $\in = 8.2\%$ and, at the same strain level, $\Delta u_f = 177$ kN/m^2 and $\bar{A} = 0.56$.

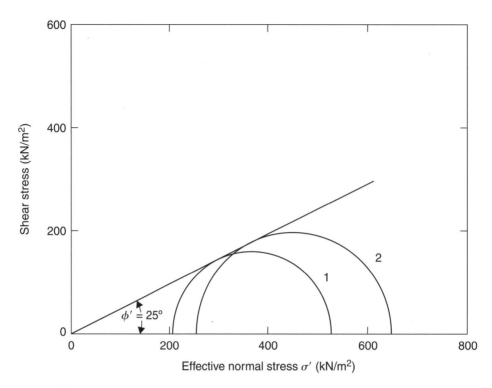

Figure 18.8. Effective-stress Mohr's circle for remolded grundite reported in Table 18.5.

12. Calculate the *effective* major and minor principal stresses at failure. The effective minor principal stress at failure is

$$\sigma_3' = \sigma_3 - \Delta u_f \qquad (18.20)$$

The effective major principal stress at failure is

$$\sigma_1' = (\sigma_3 + \Delta \sigma_f) - \Delta u_f \qquad (18.21)$$

For the test on the remolded grundite reported in Tables 18.4 and 18.5,

$$\sigma_3' = 392 - 177 = 215 \text{ kN/m}^2$$
$$\sigma_1' = (392 + 316) - 177 = 531 \text{ kN/m}^2$$

13. Collect σ_1' and σ_3' for all the specimens tested and plot Mohr's circles. Plot a failure envelope that touches Mohr's circles. The equation for the failure envelope can be given by

$$s = c' + \sigma' \tan \phi'$$

Determine the values of c' and ϕ' from the failure envelopes. Figure 18.8 shows Mohr's circles for two tests on the remolded grundite reported in Table 18.5. (*Note:* The result

Table 18.6. Range of $\bar{A}$ Values at Failure

Soil Type	$\bar{A}$ at Failure
Clays with high sensitivity	0.75–1.5
Normally consolidated clays	0.5–1.0
Overconsolidated clays	−0.5–0
Compacted sandy clay	0.5–0.75

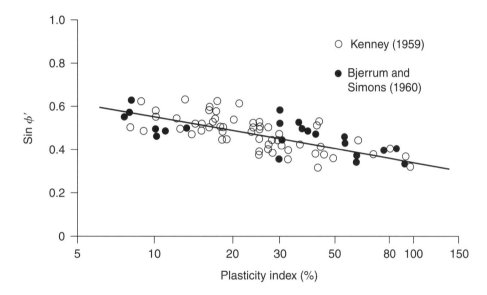

Figure 18.9. Variation of $\sin \phi'$ with plasticity index *PI* for several normally consolidated clays.

for Mohr's circle no. 2 is not given in Table 18.5.) For the failure envelope, $c' = 0$ and $\phi' = 25°$. So

$$s = \sigma' \tan 25°$$

General Comments

1. For normally consolidated soils, $c' = 0$. However, for overconsolidated soils, $c' > 0$.
2. A typical range of $\bar{A}$ values at failure for clayey soils is given in Table 18.6.
3. The friction angle ϕ' generally decreases with an increase in the plasticity index. Figure 18.9 shows the variations of $\sin \phi'$ with the plasticity index for several normally consolidated clays as reported by Kenny (1959) and Bjerrum and Simons (1960).

REFERENCES

American Society for Testing and Materials (2007). *Annual Book of ASTM Standards*, sec. 4, vol. 04.08, West Conshohocken, PA.

Atterberg, A. (1911). "Über die physikalische Bodenuntersuchung, and über die Plastizität der Tone," *Internationale Mitteilungen für Bodenkunde*, vol. 1.

Azzouz, A. S., Krizek, R. J., and Corotis, R. B. (1976). "Regression Analysis of Soil Compressibility," *Soils and Foundations*, vol. 16, no. 2, pp. 19–29.

Bjerrum, L., and Simons, N. E. (1960). "Compression of Shear Strength Characteristics of Normally Consolidated Clay," *Proceedings, Research Conference on Shear Strength of Cohesive Soils*, ASCE, pp. 711–726.

Casagrande, A. (1936). "Determination of Preconsolidation Load and Its Practical Significance," *Proceedings, First International Conference on Soil Mechanics and Foundation Engineering*, Cambridge, MA, vol. 3, pp. 60–64.

Casagrande, A. (1932). "Research of Atterberg Limits of Soils," *Public Roads*, vol. 13, no. 8, pp. 121–136.

Casagrande, A., and Fadum, R. E. (1940) "Notes on Soil Testing for Engineering Purpose," *Harvard university Graduate School of Engineering Publications no. 8.*

Das, B.M. (2006). *Principles of Geotechnical Engineering,* 6 th ed. (Thomson, Toronto, Canada).

Gurtug, Y., and Sridharan, A. (2004). "Compaction Behaviour and Prediction of Its Characteristics of Fine Grained Soils with Particular Reference to Compaction Energy," *Soils and Foundations*, vol. 44, no. 5, pp. 27–36.

Kenney, T. C. (1959). "Discussion," *Proceedings*, ASCE, vol. 85, no. SM3, pp. 67–79.

Nacci, V. A., Wang, M. C., and Demars, K. R. (1975). "Engineering Behavior of Calcareous Soils," *Proceedings, Civil Engineering in the Oceans III*, ASCE, pp. 380–400.

Nagaraj, T., and Murty, B. R. S. (1985). "Prediction of the Preconsolidation Pressure and Recompression Index of Soils," *Geotechnical Testing Journal*, vol. 8, no. 4, pp. 199–202.

Proctor, R. R. (1933). "Design and Construction of Rolled Earth Dams," *Engineering News Record*, vol. 3, pp. 245–248, 286–289, 348–351, 372–376.

Rendon–Herrero, O. (1980). "Universal Compression Index Equation," *Journal of the Geotechnical Engineering Division*, ASCE, vol. 106, no. GT11, pp. 1179–1200.

Taylor, D. W. (1942). "Research on Consolidation of Clay," *Serial No. 82*, Department of Civil and Sanitary Engineering, Massachusetts Institute of Technology, Cambridge, MA.

Terzaghi, K., and Peck, R. B. (1967). *Soil Mechanics in Engineering Pratice*, 2nd ed. (Wiley, New York)

U.S. Army Coprs of Engineers (1949). *Technical Memo 3-286*, U.S. Waterways Experiment Station, Vicksburg, MS.

Appendix A

Weight–Volume Relationships

For the weight–volume relationships given in this appendix the following notions were used:

e = void ratio
G_s = specific gravity of soil solids
n = porosity
S = degree of saturation
V = total volume of soil
V_s = volume of solids in a soil mass
V_v = volume of voids in a soil mass
V_w = volume of water in a soil mass
W = total weight of a soil mass
W_s = dry weight of a soil mass
W_w = weight of water in a soil mass
w = moisture content
γ = moist unit weight
γ_d = dry unit weight
γ_{sat} = saturated unit weight
γ_w = unit weight of water

A.1 Volume Relationships

$$e = \frac{V_v}{V_s} = \frac{n}{1-n} = \frac{G_s \gamma_w}{\gamma_d} - 1$$

$$n = \frac{V_v}{V} = \frac{e}{1+e} = 1 - \frac{\gamma_d}{G_s \gamma_w}$$

$$S = \frac{V_w}{V_v} = \frac{w G_s}{e}$$

A.2 Weight Relationships

Moisture content

$$w = \frac{W_w}{W_s} = \frac{Se}{G_s}$$

Moist unit weight

$$\gamma = \frac{W}{V}$$

$$\gamma = \frac{(1+w)G_s \gamma_w}{1+e}$$

$$\gamma = \frac{(G_s + Se)\gamma_w}{1+e}$$

$$\gamma = \frac{(1+w)G_s\gamma_w}{1+(wG_s/S)}$$

$$\gamma = G_s\gamma_w(1-n)(1+w)$$

$$\gamma = G_s\gamma_w(1-n) + nS\gamma_w$$

Dry unit weight

$$\gamma_d = \frac{W_s}{V}$$

$$\gamma_d = \frac{\gamma}{1+w}$$

$$\gamma_d = \frac{G_s\gamma_w}{1+e}$$

$$\gamma_d = G_s\gamma_w(1-n)$$

$$\gamma_d = \frac{G_s\gamma_w}{1+(wG_s/S)}$$

$$\gamma_d = \frac{eS\gamma_w}{(1+e)w}$$

$$\gamma_d = \gamma_{\text{sat}} - \frac{e\gamma_w}{1+e}$$

$$\gamma_d = \gamma_{\text{sat}} - n\gamma_w$$

$$\gamma_d = \frac{(\gamma_{\text{sat}} - \gamma_w)G_s}{G_s - 1}$$

Saturated unit weight

$$\gamma_{\text{sat}} = \frac{(G_s + e)\gamma_w}{1+e}$$

$$\gamma_{\text{sat}} = [(1-n)G_s + n]\gamma_w$$

$$\gamma_{\text{sat}} = \frac{1+w_{\text{sat}}}{1+w_{\text{sat}}G_s}G_s\gamma_w$$

$$\gamma_{\text{sat}} = \frac{e}{w_{\text{sat}}}\frac{1+w_{\text{sat}}}{1+e}\gamma_w$$

$$\gamma_{\text{sat}} = n\frac{1+w_{\text{sat}}}{w_{\text{sat}}}\gamma_w$$

$$\gamma_{\text{sat}} = \gamma_d + \frac{e}{1+e}\gamma_w$$

$$\gamma_{\text{sat}} = \gamma_d + n\gamma_w$$

$$\gamma_{\text{sat}} = \left(1 - \frac{1}{G_s}\right)\gamma_d + \gamma_w$$

$$\gamma_{\text{sat}} = (1 + w_{\text{sat}})\gamma_d$$

Appendix B

Conversion Factors

Table B.1. Conversion from English to SI Units

Quantity	English	SI
Length	1 in.	25.4 mm
	1 ft	0.3048 m
		304.8 mm
Area	1 in.^{2}	6.4516×10^{-4} m^2
		6.4516 cm^2
		645.16 mm^2
	1 ft^2	929×10^{-4} m^2
		929.03 cm^2
		92903 mm^2
Volume	1 in.^{3}	16.387 cm^3
	1 ft^3	0.028317 m^3
		28317 mm^3
Velocity	1 ft/s	304.8 mm/s
		0.3048 m/s
	1 ft/min	5.08 mm/s
		0.00508 m/s
Force	1 lb	4.448 N
Stress	1 lb/in.^{2}	6.9 kN/m^2
	1 lb/ft^2	47.88 N/m^2
Unit weight	1 lb/ft^3	157.06 N/m^3
Coefficient of consolidation	1 in.^{2}/s	6.452 cm^2/s
	1 ft^2/s	929.03 cm^2/s

Table B.2. Conversion from SI to English Units

Quantity	English	SI
Length	1 mm	3.937×10^{-2} in.
		3.218×10^{-3} ft
	1 m	39.37 in.
		3.281 ft
Area	1 cm²	0.155 in.²
		1.076×10^{-3} ft²
	1 m²	1550 in.²
		10.76 ft²
Volume	1 cm³	0.061 in.³
		3.531×10^{-5} ft³
	1 m³	61023.74 in.³
		35.315 ft³
Velocity	1 cm/s	1.969 ft/min
		1034643.6 ft/year
Force	1 N	0.22482 lb
	1 kN	0.22482 kip
Stress	1 kN/m²	0.145 lb/in.²
		20.89 lb/ft²
Unit weight	1 kN/m³	6.367 lb/ft³
Coefficient of consolidation	1 cm²/s	0.155 in.²/s
		2.883×10^3 ft²/month
Mass	1 kg	2.2046 lb
		2.2046×10^{-3} kip

APPENDIX C

Data Sheets for Laboratory Experiments

Determination of Water Content

Description of soil _____ Sample no. _____

Location _____

Tested by _____ Date _____

Item	Test No.		
	1	2	3
Can no.			
Mass of can, M_1 (g)			
Mass of can + wet soil, M_2 (g)			
Mass of can + dry soil, M_3 (g)			
Mass of moisture, $M_2 - M_3$ (g)			
Mass of dry soil, $M_3 - M_1$ (g)			
Water content, w (%) $= \dfrac{M_2 - M_3}{M_3 - M_1} \times 100$			

Average water content w _____ %

Specific Gravity of Soil Solids

Description of soil _____ Sample no. _____

Volume of flask at 20°C _____ ml   Temperature of test _____ °C A _____ (Table 3.4)

Location _____

Tested by _____ Date _____

Item	Test No.		
	1	2	3
Volumetric flask no.			
Mass of flask + water filled to mark, M_1 (g)			
Mass of flask + soil + water filled to mark, M_2 (g)			
Mass of dry soil, M_s (g)			
Mass of equal volume of water and soil solids, M_w (g) $= (M_1 + M_s) - M_2$			
$G_{s(\text{at } T_1°C)} = M_s/M_w$			
$G_{s(\text{at } 20°C)} = G_{s(\text{at } T_1°C)} \times A$			

Average G_s _____

175

Sieve Analysis

Description of soil _____ Sample no. _____

Mass of oven-dry specimen M _____ g

Location _____

Tested by _____ Date _____

Sieve No.	Sieve Opening (mm)	Mass of Soil Retained on Each Sieve M_n (g)	Percent of Mass Retained on Each Sieve R_n	Cumulative Percent Retained $\sum R_n$	Percent Finer $100 - \sum R_n$
Pan	_____				

$$\sum \text{_____} = M_1$$

Mass loss during sieve analysis: $\dfrac{M - M_1}{M} \times 100 =$ _____ % (OK if less than 2%)

Hydrometer Analysis

Description of soil _____ Sample no. _____

Location _____

G_s _____ Hydrometer type _____

Dry mass of soil M_s _____ g Temperature of test T _____°C

Meniscus correction F_m _____ Zero correction F_s _____

Temperature correction F_T _____ [Eq. (5.6)]

Tested by _____ Date _____

Time (min) (1)	Hydrometer Reading R (2)	R_{cp} (3)	Percent Finer, $\frac{a*R_{cp}}{50} \times 100$ (4)	R_{cL} (5)	$L^\dagger$ (cm) (6)	$A^\ddagger$ (7)	D (mm) (8)

*Table 5.4; †Table 5.1; ‡Table 5.2.

Liquid Limit Test

Description of soil _____ Sample no. _____

Location _____

Tested by _____ Date _____

Test No.	1	2	3
Can no.			
Mass of can, M_1 (g)			
Mass of can + moist soil, M_2 (g)			
Mass of can + dry soil, M_3 (g)			
Moisture content, $w\ (\%) = \dfrac{M_2 - M_3}{M_3 - M_1} \times 100$			
Number of blows, N			

Liquid limit LL _____

Flow index F_I _____

Plastic Limit Test

Description of soil _____ Sample no. _____

Location _____

Tested by _____ Date _____

Test No.	1	2	3
Can no.			
Mass of can, M_1 (g)			
Mass of can + moist soil, M_2 (g)			
Mass of can + dry soil, M_3 (g)			
$PL = \dfrac{M_2 - M_3}{M_3 - M_1} \times 100$			

Plasticity index $PI = LL - PL =$ _____ $=$ _____

Shrinkage Limit Test

Description of soil _____ Sample no. _____

Location _____

Tested by _____ Date _____

Test No.		
Mass of coated shrinkage limit dish, M_1 (g)		
Mass of dish + wet soil, M_2 (g)		
Mass of dish + dry soil, M_3 (g)		
$w_i\ (\%) = \dfrac{M_2 - M_3}{M_3 - M_1} \times 100$		
Mass of mercury to fill dish, M_4 (g)		
Mass of mercury displaced by soil pat, M_5 (g)		
$\Delta w_i\ (\%) = \dfrac{M_4 - M_5}{13.6\,(M_3 - M_1)} \times 100$		
$SL = w_i - \dfrac{M_4 - M_5}{13.6\,(M_3 - M_1)} \times 100$		

Constant-Head Permeability Test

Determination of Void Ratio of Specimen

Description of soil _____ Sample no. _____

Location _____

Length of specimen L _____ cm   Diameter of specimen D _____ cm

Tested by _____ Date _____

Volume of specimen, $V = \dfrac{\pi}{4}D^2L$ (cm^3)	
Specific gravity of soil solids, G_s	
Mass of specimen tube with fittings, M_1 (g)	
Mass of tube with fittings and specimen, M_2 (g)	
Dry density of specimen, $\rho_d = \dfrac{M_2 - M_1}{V}$ (g/cm^3)	

Void ratio of specimen $e = \dfrac{G_s\rho_w}{\rho_d} - 1 =$ _____

(*Note:* $\rho_w = 1$ g/cm^3)

Constant-Head Permeability Test
Determination of Coefficient of Permeability

Description of soil _____ Sample no. _____

Location _____

Length of specimen L _____ cm   Diameter of specimen D _____ cm

Tested by _____ Date _____

Test No.	1	2	3
Average flow, Q (cm^3)			
Time of collection, t (s)			
Temperature of water, $T(°C)$			
Head difference, h (cm)			
Diameter of specimen, D (cm)			
Length of specimen, L (cm)			
Area of specimen, $A = \dfrac{\pi}{4}D^2$ (cm^2)			
$k = \dfrac{QL}{Aht}$ (cm/s)			

Average $k =$ _____ cm/s

$$k_{20°C} = k_{T°C}\frac{\eta_{T°C}}{\eta_{20°C}} = \text{_____} = \text{_____} \text{ cm/s}$$

Falling-Head Permeability Test
Determination of Void Ratio of Specimen

Description of soil _____ Sample no. _____

Location _____

Length of specimen L _____ cm   Diameter of specimen D _____ cm

Tested by _____ Date _____

Volume of specimen, $V = \dfrac{\pi}{4}D^2 L \ (\text{cm}^3)$	
Specific gravity of soil solids, G_s	
Mass of specimen tube with fittings, M_1 (g)	
Mass of tube with fittings and specimen, M_2 (g)	
Dry density of specimen, $\rho_d = \dfrac{M_2 - M_1}{V} \ (\text{g/cm}^3)$	

Void ratio of specimen $e = \dfrac{G_s \rho_w}{\rho_d} - 1 = \underline{\hspace{2cm}}$

(*Note:* $\rho_w = 1 \text{ g/cm}^3$)

Falling-Head Permeability Test
Determination of Coefficient of Permeability

Description of soil _____ Sample no. _____

Location _____

Length of specimen L _____ cm   Diameter of specimen D _____ cm

Tested by _____ Date _____

Test No.	1	2	3
Diameter of specimen, D (cm)			
Length of specimen, L (cm)			
Area of specimen, A (cm^2)			
Beginning head difference, h_1 (cm)			
Ending head difference, h_2 (cm)			
Test duration, t (s)			
Volume of water flow through specimen, V_w (cm^3)			
$k = \dfrac{2.303 V_w L}{(h_1 - h_2)tA} \log \dfrac{h_1}{h_2}$ (cm/s)			

Average $k =$ _____ cm/s

$$k_{20°C} = k_{T°C} \frac{\eta_{T°C}}{\eta_{20°C}} = \underline{\hspace{2cm}} = \underline{\hspace{2cm}} \text{ cm/s}$$

Standard Proctor Compaction Test
Determination of Dry Unit Weight

Description of soil _____ Sample no. _____

Location _____

Volume of mold _____ ft^3 Weight of hammer _____ lb

Number of blows/layer _____   Number of layers _____

Tested by _____   Date _____

Test	1	2	3	4	5	6
1. Weight of mold and base plate, W_1 (lb)						
2. Weight of mold and base plate + moist soil, W_2 (lb)						
3. Weight of moist soil, $W_2 - W_1$ (lb)						
4. Moist unit weight, $\gamma = \dfrac{W_2 - W_1}{1/30}$ (lb/ft^3)						
5. Moisture can number						
6. Mass of moisture can, M_3 (g)						
7. Mass of can + moist soil, M_4 (g)						
8. Mass of can + dry soil, M_5 (g)						
9. Moisture content, $w\ (\%) = \dfrac{M_4 - M_5}{M_5 - M_3} \times 100$						
10. Dry unit weight of compaction, γ_d (lb/ft^3) = $\dfrac{\gamma}{1 + (w\ (\%)/100)}$						

Standard Proctor Compaction Test

Zero-Air-Void Unit Weight

Description of soil _____ Sample no. _____

Location _____

Tested by _____ Date _____

Specific Gravity of Soil Solids G_s	Assumed Moisture Content w (%)	Unit Weight of Water γ_w (lb/ft³)	γ_{zav}^* (lb/ft³)

*Equation (12.1).

Modified Proctor Compaction Test
Determination of Dry Unit Weight

Description of soil _____ Sample no. _____

Location _____

Volume of mold _____ ft^3 Weight of hammer _____ lb

Number of blows/layer _____   Number of layers _____

Tested by _____   Date _____

Test	1	2	3	4	5	6
1. Weight of mold and base plate, W_1 (lb)						
2. Weight of mold and base plate + moist soil, W_2 (lb)						
3. Weight of moist soil, $W_2 - W_1$ (lb)						
4. Moist unit weight, $\gamma = \dfrac{W_2 - W_1}{1/30}$ (lb/ft^3)						
5. Moisture can number						
6. Mass of moisture can, M_3 (g)						
7. Mass of can + moist soil, M_4 (g)						
8. Mass of can + dry soil, M_5 (g)						
9. Moisture content, $w\ (\%) = \dfrac{M_4 - M_5}{M_5 - M_3} \times 100$						
10. Dry unit weight of compaction, γ_d (lb/ft^3) = $\dfrac{\gamma}{1 + (w\ (\%)/100)}$						

Modified Proctor Compaction Test

Zero-Air-Void Unit Weight

Description of soil _____ Sample no. _____

Location _____

Tested by _____ Date _____

Specific Gravity of Soil Solids G_s	Assumed Moisture Content w (%)	Unit Weight of Water γ_w (lb/ft^3)	γ^*_{zav} (lb/ft^3)

*Equation (12.1).

Sand Cone Method
Field Unit Weight

Location _____

Tested by _____ Date _____

Item	Quantity
Calibration of Unit Weight of Ottawa Sand	
1. Weight of Proctor mold, W_1	
2. Weight of Proctor mold + sand, W_2	
3. Volume of mold, V_1	
4. Dry unit weight, $\gamma_{d(\text{sand})} = \dfrac{W_2 - W_1}{V_1}$	
Calibration Cone	
5. Weight of bottle + cone + sand (before use), W_3	
6. Weight of bottle + cone + sand (after use), W_4	
7. Weight of sand to fill cone, $W_c = W_4 - W_3$	
Results from Field Tests	
8. Weight of bottle + cone + sand (before use), W_6	
9. Weight of bottle + cone + sand (after use), W_8	
10. Volume of hole, $V_2 = \dfrac{W_6 - W_8 - W_c}{\gamma_{d(\text{sand})}}$	
11. Weight of gallon can, W_5	
12. Weight of gallon can + moist soil, W_7	
13. Weight of gallon can + dry soil, W_9	
14. Moist unit weight of soil in field, $\gamma = \dfrac{W_7 - W_5}{V_2}$	
15. Moisture content in field, $w\,(\%) = \dfrac{W_7 - W_9}{W_9 - W_5} \times 100$	
16. Dry unit weight in field, $\gamma_{d(\text{sand})} = \dfrac{\gamma}{1 + (w\,(\%)/100)}$	

Direct Shear Test on Sand

Void Ratio Calculation

Description of soil _____ Sample no. _____

Location _____

Tested by _____ Date _____

Item	Quantity
1. Specimen length, L (in.)	
2. Specimen width, B (in.)	
3. Specimen height, H (in.)	
4. Mass of porcelain dish + dry sand (before use), M_1 (g)	
5. Mass of porcelain dish + dry sand (after use), M_2 (g)	
6. Dry unit weight of specimen, $\gamma_d \, (\text{lb/ft}^3) = \dfrac{M_1 - M_2}{LBH \, (\text{in.}^3)} \times 3.808$	
7. Specific gravity of soil solids, G_s	
8. Void ratio, $e = \dfrac{G_s \gamma_w}{\gamma_d} - 1$ (*Note:* $\gamma_w = 62.4 \, \text{lb/ft}^3$; γ_d is in lb/ft^3)	

Direct Shear Test on Sand
Stress and Displacement Calculation

Description of soil _____   Sample no. _____

Location _____

Normal load N _____ lb   Void ratio e _____

Tested by _____   Date _____

Normal Stress σ' (lb/in.^{2}) (1)	Horizontal Displacement (in.) (2)	Vertical Displacement* (in.) (3)	No. of div. in Proving Ring Dial Gauge (4)	Proving Ring Calibration Factor (lb/div.) (5)	Shear Force S (lb) (6)	Shear Stress τ (lb/in.^{2}) (7)

* Plus (+) sign means expansion.

Direct Shear Test on Sand
Stress and Displacement Calculation

Description of soil _____ Sample no. _____

Location _____

Normal load N _____ lb   Void ratio e _____

Tested by _____ Date _____

Normal Stress σ' (lb/in.^{2}) (1)	Horizontal Displacement (in.) (2)	Vertical Displacement* (in.) (3)	No. of div. in Proving Ring Dial Gauge (4)	Proving Ring Calibration Factor (lb/div.) (5)	Shear Force S (lb) (6)	Shear Stress τ (lb/in.^{2}) (7)

* Plus (+) sign means expansion.

Direct Shear Test on Sand

Stress and Displacement Calculation

Description of soil _____ Sample no. _____

Location _____

Normal load N _____ lb   Void ratio e _____

Tested by _____ Date _____

Normal Stress σ' (lb/in.^{2}) (1)	Horizontal Displacement (in.) (2)	Vertical Displacement* (in.) (3)	No. of div. in Proving Ring Dial Gauge (4)	Proving Ring Calibration Factor (lb/div.) (5)	Shear Force S (lb) (6)	Shear Stress τ (lb/in.^{2}) (7)

* Plus (+) sign means expansion.

Unconfined Compression Test

Description of soil _____ Specimen no. _____

Location _____

Moist mass of specimen _____ g   Moisture content _____ %

Length of specimen L _____ in.   Diameter of specimen _____ in.

Proving ring calibration factor: 1 div. = _____ lb   Area $A_0 = \frac{\Pi}{4}D^2 =$ _____ in.^{2}

Tested by _____   Date _____

Specimen Deformation ΔL (in.) (1)	Vertical Strain $\varepsilon = \frac{\Delta L}{L}$ (2)	Proving Ring Dial Reading (no. of small div.) (3)	Load P (Column 3 × calibration factor) (lb) (4)	Corrected Area $A_c = \frac{A_0}{1-\varepsilon}$ (in.^{2}) (5)	Stress $\sigma = \frac{\text{Column 4}}{\text{Column 5}}$ (lb/in.^{2}) (6)

Consolidation Test
Time versus Vertical Dial Reading

Description of soil _____

Location _____

Tested by _____ Date _____

Pressure on specimen _____ ton/ft^2 Pressure on specimen _____ ton/ft^2

Time after Load Application t (min)	$\sqrt{t}$ (min)$^{0.5}$	Vertical Dial Reading (in)	Time after Load Application t (min)	$\sqrt{t}$ (min)$^{0.5}$	Vertical Dial Reading (in.)

Consolidation Test
Time versus Vertical Dial Reading

Description of soil _____

Location _____

Tested by _____ Date _____

Pressure on specimen _____ ton/ft^2 Pressure on specimen _____ ton/ft^2

Time after Load Application t (min)	$\sqrt{t}$ (min)$^{0.5}$	Vertical Dial Reading (in)	Time after Load Application t (min)	$\sqrt{t}$ (min)$^{0.5}$	Vertical Dial Reading (in.)

Consolidation Test
Time versus Vertical Dial Reading

Description of soil _____

Location _____

Tested by _____ Date _____

Pressure on specimen _____ ton/ft^2 Pressure on specimen _____ ton/ft^2

Time after Load Application t (min)	$\sqrt{t}$ (min)$^{0.5}$	Vertical Dial Reading (in)	Time after Load Application t (min)	$\sqrt{t}$ (min)$^{0.5}$	Vertical Dial Reading (in.)

Consolidation Test

Pressure, Void Ratio, and Calculation of Coefficient of Consolidation

Description of soil _____ Location _____

Specimen diameter _____ in.   Initial specimen height $H_{t(i)}$ _____ in

Height of solids H_s _____ cm = _____ in.

Moisture content: Beginning of test _____ %   End of test _____ %

Mass of dry soil specimen _____ g G_s _____

Tested by _____ Date _____

Pressure p (ton/ft^2)	Final Dial Reading (in.)	Change in Specimen Height ΔH (in.)	Final Specimen Height $H_{t(f)}$ (in.)	Height of Void H_v (in.)	Final void ratio e	Average Height during Consolidation $H_{t(av)}$ (in.)	Fitting Time (s)		$C_v \times 10^3$ (in.^{2}/sec) from	
							t_{90}	t_{50}	t_{90}	t_{50}
(1)	(2)	(3)	(4)	(5)	(6)	(7)	(8)	(9)	(10)	(11)

Unconsolidated-Undrained Triaxial Test

Preliminary Data

Description of soil _____ Specimen no. _____

Location _____

Tested by _____ Date _____

Item	Quantity
1. Moist mass of specimen (end of test), M_1	
2. Dry mass of specimen, M_2	
3. Moisture content (end of test), $w\ (\%) = \dfrac{M_1 - M_2}{M_2} \times 100$	
4. Initial average length of specimen, L_0	
5. Initial average diameter of specimen, D_0	
6. Initial area, $A_0 = \dfrac{\pi}{4} D_0^2$	
7. Specific gravity of soil solids, G_s	
8. Final degree of saturation	
9. Cell confining pressure, σ_3	
10. Proving ring calibration factor	

Unconsolidated-Undrained Triaxial Test

Axial Stress–Strain Calculation

Specimen Deformation ΔL (in.) (1)	Vertical Strain $\varepsilon = \dfrac{\Delta L}{L_0}$ (2)	Proving Ring Dial Reading (no. of small div.) (3)	Piston Load P (Column 3 × calibration factor) (lb) (4)	Corrected Area $A = \dfrac{A_0}{1-\varepsilon}$ (in.^{2}) (5)	Deviatory Stress $\Delta\sigma = \dfrac{P}{A}$ (lb/in.^{2}) (6)

Consolidated-Undrained Triaxial Test

Preliminary Data

Description of soil _____ Specimen no. _____

Location _____

Tested by _____ Date _____

Beginning of Test	
1. Moist unit weight of specimen (beginning of test)	
2. Moisture content (beginning of test)	
3. Initial length of specimen, L_0	
4. Initial diameter of specimen, D_0	
5. Initial area of specimen, $A_0 = \dfrac{\pi}{4}D_0^2$	
6. Initial volume of specimen, $V_0 = A_0 L_0$	
After Consolidation of Saturated Specimen	
7. Cell consolidation pressure, σ_3	
8. Net drainage from specimen during consolidation, ΔV	
9. Volume of specimen after consolidation, $V_0 - \Delta V = V_c$	
10. Length of specimen after consolidation, $L_c = L_0 \left(\dfrac{V_c}{V_0}\right)^{1/3}$	
11. Area of specimen after consolidation, $A_c = A_0 \left(\dfrac{V_c}{V_0}\right)^{2/3}$	

Consolidated-Undrained Triaxial Test

Axial Stress–Strain Calculation

Proving ring calibration factor _____

Specimen Deformation ΔL (cm) (1)	Vertical Strain $\varepsilon = \frac{\Delta L}{L_c}$ (2)	Proving Ring Dial Reading (no. of small div.) (3)	Piston Load P (N) (4)	Corrected Area $A = \frac{A_c}{1-\varepsilon}$ (cm^2) (5)	Deviatory Stress $\Delta\sigma = \frac{P}{A}$ (kN/m^2) (6)	Excess Pore-Water Pressure Δu (kN/m^2) (7)	$\bar{A} = \frac{\Delta u}{\Delta\sigma}$ (8)

Consolidated-Undrained Triaxial Test

Preliminary Data

Description of soil _____ Specimen no. _____

Location _____

Tested by _____ Date _____

Beginning of Test	
1. Moist unit weight of specimen (beginning of test)	
2. Moisture content (beginning of test)	
3. Initial length of specimen, L_0	
4. Initial diameter of specimen, D_0	
5. Initial area of specimen, $A_0 = \dfrac{\pi}{4}D_0^2$	
6. Initial volume of specimen, $V_0 = A_0 L_0$	
After Consolidation of Saturated Specimen	
7. Cell consolidation pressure, σ_3	
8. Net drainage from specimen during consolidation, ΔV	
9. Volume of specimen after consolidation, $V_0 - \Delta V = V_c$	
10. Length of specimen after consolidation, $L_c = L_0\left(\dfrac{V_c}{V_0}\right)^{1/3}$	
11. Area of specimen after consolidation, $A_c = A_0\left(\dfrac{V_c}{V_0}\right)^{2/3}$	

Consolidated-Undrained Triaxial Test

Axial Stress–Strain Calculation

Proving ring calibration factor _____

Specimen Deformation ΔL (cm) (1)	Vertical Strain $\varepsilon = \frac{\Delta L}{L_c}$ (2)	Proving Ring Dial Reading (no. of small div.) (3)	Piston Load P (N) (4)	Corrected Area $A = \frac{A_c}{1-\varepsilon}$ (cm^2) (5)	Deviatory Stress $\Delta\sigma = \frac{P}{A}$ (kN/m^2) (6)	Excess Pore-Water Pressure Δu (kN/m^2) (7)	$\bar{A} = \frac{\Delta u}{\Delta\sigma}$ (8)

APPENDIX D

Data Sheets for Preparation of Laboratory Reports

Determination of Water Content

Description of soil _____ Sample no. _____

Location _____

Tested by _____ Date _____

Item	Test No.		
	1	**2**	**3**
Can no.			
Mass of can, M_1 (g)			
Mass of can + wet soil, M_2 (g)			
Mass of can + dry soil, M_3 (g)			
Mass of moisture, $M_2 - M_3$ (g)			
Mass of dry soil, $M_3 - M_1$ (g)			
Water content, w (%) $= \dfrac{M_2 - M_3}{M_3 - M_1} \times 100$			

Average water content w _____ %

Specific Gravity of Soil Solids

Description of soil _____ Sample no. _____

Volume of flask at 20°C _____ ml Temperature of test _____ °C A _____ (Table 3.4)

Location _____

Tested by _____ Date _____

Item	Test No.		
	1	**2**	**3**
Volumetric flask no.			
Mass of flask + water filled to mark, M_1 (g)			
Mass of flask + soil + water filled to mark, M_2 (g)			
Mass of dry soil, M_s (g)			
Mass of equal volume of water and soil solids, M_w (g) $= (M_1 + M_s) - M_2$			
$G_{s(\text{at } T_1°C)} = M_s/M_w$			
$G_{s(\text{at } 20°C)} = G_{s(\text{at } T_1°C)} \times A$			

Average G_s _____

Sieve Analysis

Description of soil _____ Sample no. _____

Mass of oven-dry specimen M _____ g

Location _____

Tested by _____ Date _____

Sieve No.	Sieve Opening (mm)	Mass of Soil Retained on Each Sieve M_n (g)	Percent of Mass Retained on Each Sieve R_n	Cumulative Percent Retained $\sum R_n$	Percent Finer $100 - \sum R_n$
Pan	____				

$\sum$ _____ $= M_1$

Mass loss during sieve analysis: $\dfrac{M - M_1}{M} \times 100 =$ _____ % (OK if less than 2%)

Hydrometer Analysis

Description of soil _____ Sample no. _____

Location _____

G_s _____ Hydrometer type _____

Dry mass of soil M_s _____ g Temperature of test T _____ °C

Meniscus correction F_m _____ Zero correction F_s _____

Temperature correction F_T _____ [Eq. (5.6)]

Tested by _____ Date _____

Time (min) (1)	Hydrometer Reading R (2)	R_{cp} (3)	Percent Finer, $\frac{a*R_{cp}}{50} \times 100$ (4)	R_{cL} (5)	$L^{\dagger}$ (cm) (6)	$A^{\ddagger}$ (7)	D (mm) (8)

*Table 5.4; †Table 5.1; ‡Table 5.2.

Liquid Limit Test

Description of soil _____ Sample no. _____

Location _____

Tested by _____ Date _____

Test No.	1	2	3
Can no.			
Mass of can, M_1 (g)			
Mass of can + moist soil, M_2 (g)			
Mass of can + dry soil, M_3 (g)			
Moisture content, $w\,(\%) = \dfrac{M_2 - M_3}{M_3 - M_1} \times 100$			
Number of blows, N			

Liquid limit LL _____

Flow index F_I _____

Plastic Limit Test

Description of soil _____ Sample no. _____

Location _____

Tested by _____ Date _____

Test No.	1	2	3
Can no.			
Mass of can, M_1 (g)			
Mass of can + moist soil, M_2 (g)			
Mass of can + dry soil, M_3 (g)			
$PL = \dfrac{M_2 - M_3}{M_3 - M_1} \times 100$			

Plasticity index $PI = LL - PL =$ _____ $=$ _____

Shrinkage Limit Test

Description of soil _____ Sample no. _____

Location _____

Tested by _____ Date _____

Test No.		
Mass of coated shrinkage limit dish, M_1 (g)		
Mass of dish + wet soil, M_2 (g)		
Mass of dish + dry soil, M_3 (g)		
$w_i\,(\%) = \dfrac{M_2 - M_3}{M_3 - M_1} \times 100$		
Mass of mercury to fill dish, M_4 (g)		
Mass of mercury displaced by soil pat, M_5 (g)		
$\Delta w_i\,(\%) = \dfrac{M_4 - M_5}{13.6\,(M_3 - M_1)} \times 100$		
$SL = w_i - \dfrac{M_4 - M_5}{13.6\,(M_3 - M_1)} \times 100$		

Constant-Head Permeability Test
Determination of Void Ratio of Specimen

Description of soil _____ Sample no. _____

Location _____

Length of specimen L _____ cm   Diameter of specimen D _____ cm

Tested by _____ Date _____

Volume of specimen, $V = \dfrac{\pi}{4}D^2L$ (cm^3)	
Specific gravity of soil solids, G_s	
Mass of specimen tube with fittings, M_1 (g)	
Mass of tube with fittings and specimen, M_2 (g)	
Dry density of specimen, $\rho_d = \dfrac{M_2 - M_1}{V}$ (g/cm^3)	

Void ratio of specimen $e = \dfrac{G_s\rho_w}{\rho_d} - 1 = $ _____

(*Note:* $\rho_w = 1$ g/cm^3)

Constant-Head Permeability Test
Determination of Coefficient of Permeability

Description of soil _____ Sample no. _____

Location _____

Length of specimen L _____ cm Diameter of specimen D _____ cm

Tested by _____ Date _____

Test No.	1	2	3
Average flow, Q (cm^3)			
Time of collection, t (s)			
Temperature of water, $T(°C)$			
Head difference, h (cm)			
Diameter of specimen, D (cm)			
Length of specimen, L (cm)			
Area of specimen, $A = \dfrac{\pi}{4}D^2$ (cm^2)			
$k = \dfrac{QL}{Aht}$ (cm/s)			

Average $k =$ _____ cm/s

$k_{20°C} = k_{T°C}\dfrac{\eta_{T°C}}{\eta_{20°C}} =$ _____ $=$ _____ cm/s

Falling-Head Permeability Test
Determination of Void Ratio of Specimen

Description of soil _____ Sample no. _____

Location _____

Length of specimen L _____ cm   Diameter of specimen D _____ cm

Tested by _____ Date _____

Volume of specimen, $V = \dfrac{\pi}{4}D^2L \ (\text{cm}^3)$	
Specific gravity of soil solids, G_s	
Mass of specimen tube with fittings, M_1 (g)	
Mass of tube with fittings and specimen, M_2 (g)	
Dry density of specimen, $\rho_d = \dfrac{M_2 - M_1}{V} \ (\text{g/cm}^3)$	

Void ratio of specimen $e = \dfrac{G_s\rho_w}{\rho_d} - 1 = $ _____

(*Note:* $\rho_w = 1 \ \text{g/cm}^3$)

Falling-Head Permeability Test
Determination of Coefficient of Permeability

Description of soil _____ Sample no. _____

Location _____

Length of specimen L _____ cm Diameter of specimen D _____ cm

Tested by _____ Date _____

Test No.	1	2	3
Diameter of specimen, D (cm)			
Length of specimen, L (cm)			
Area of specimen, A (cm^2)			
Beginning head difference, h_1 (cm)			
Ending head difference, h_2 (cm)			
Test duration, t (s)			
Volume of water flow through specimen, V_w (cm^3)			
$k = \dfrac{2.303 V_w L}{(h_1 - h_2)tA} \log \dfrac{h_1}{h_2}$ (cm/s)			

Average $k =$ _____ cm/s

$$k_{20°C} = k_{T°C} \frac{\eta_{T°C}}{\eta_{20°C}} = \text{_____} = \text{_____ cm/s}$$

Standard Proctor Compaction Test
Determination of Dry Unit Weight

Description of soil _____ Sample no. _____

Location _____

Volume of mold _____ ft^3 Weight of hammer _____ lb

Number of blows/layer _____ Number of layers _____

Tested by _____ Date _____

Test	1	2	3	4	5	6
1. Weight of mold and base plate, W_1 (lb)						
2. Weight of mold and base plate + moist soil, W_2 (lb)						
3. Weight of moist soil, $W_2 - W_1$ (lb)						
4. Moist unit weight, $\gamma = \dfrac{W_2 - W_1}{1/30}$ (lb/ft^3)						
5. Moisture can number						
6. Mass of moisture can, M_3 (g)						
7. Mass of can + moist soil, M_4 (g)						
8. Mass of can + dry soil, M_5 (g)						
9. Moisture content, $w\,(\%) = \dfrac{M_4 - M_5}{M_5 - M_3} \times 100$						
10. Dry unit weight of compaction, γ_d (lb/ft^3) $= \dfrac{\gamma}{1 + (w\,(\%)/100)}$						

Standard Proctor Compaction Test

Zero-Air-Void Unit Weight

Description of soil _____ Sample no. _____

Location _____

Tested by _____ Date _____

Specific Gravity of Soil Solids G_s	Assumed Moisture Content w (%)	Unit Weight of Water γ_w (lb/ft^3)	γ_{zav}^* (lb/ft^3)

*Equation (12.1).

Modified Proctor Compaction Test
Determination of Dry Unit Weight

Description of soil _____ Sample no. _____

Location _____

Volume of mold _____ ft^3 Weight of hammer _____ lb

Number of blows/layer _____ Number of layers _____

Tested by _____ Date _____

Test	1	2	3	4	5	6
1. Weight of mold and base plate, W_1 (lb)						
2. Weight of mold and base plate + moist soil, W_2 (lb)						
3. Weight of moist soil, $W_2 - W_1$ (lb)						
4. Moist unit weight, $\gamma = \dfrac{W_2 - W_1}{1/30}$ (lb/ft^3)						
5. Moisture can number						
6. Mass of moisture can, M_3 (g)						
7. Mass of can + moist soil, M_4 (g)						
8. Mass of can + dry soil, M_5 (g)						
9. Moisture content, $w\,(\%) = \dfrac{M_4 - M_5}{M_5 - M_3} \times 100$						
10. Dry unit weight of compaction, γ_d (lb/ft^3) $= \dfrac{\gamma}{1 + (w\,(\%)/100)}$						

Modified Proctor Compaction Test

Zero-Air-Void Unit Weight

Description of soil _____ Sample no. _____

Location _____

Tested by _____ Date _____

Specific Gravity of Soil Solids G_s	Assumed Moisture Content w (%)	Unit Weight of Water γ_w (lb/ft^3)	γ_{zav}^* (lb/ft^3)

*Equation (12.1).

Sand Cone Method

Field Unit Weight

Location _____

Tested by _____ Date _____

Item	Quantity
Calibration of Unit Weight of Ottawa Sand	
1. Weight of Proctor mold, W_1	
2. Weight of Proctor mold + sand, W_2	
3. Volume of mold, V_1	
4. Dry unit weight, $\gamma_{d\,(\text{sand})} = \dfrac{W_2 - W_1}{V_1}$	
Calibration Cone	
5. Weight of bottle + cone + sand (before use), W_3	
6. Weight of bottle + cone + sand (after use), W_4	
7. Weight of sand to fill cone, $W_c = W_4 - W_3$	
Results from Field Tests	
8. Weight of bottle + cone + sand (before use), W_6	
9. Weight of bottle + cone + sand (after use), W_8	
10. Volume of hole, $V_2 = \dfrac{W_6 - W_8 - W_c}{\gamma_{d\,(\text{sand})}}$	
11. Weight of gallon can, W_5	
12. Weight of gallon can + moist soil, W_7	
13. Weight of gallon can + dry soil, W_9	
14. Moist unit weight of soil in field, $\gamma = \dfrac{W_7 - W_5}{V_2}$	
15. Moisture content in field, $w\,(\%) = \dfrac{W_7 - W_9}{W_9 - W_5} \times 100$	
16. Dry unit weight in field, $\gamma_{d\,(\text{sand})} = \dfrac{\gamma}{1 + (w\,(\%)/100)}$	

Direct Shear Test on Sand

Void Ratio Calculation

Description of soil _____ Sample no. _____

Location _____

Tested by _____ Date _____

Item	Quantity
1. Specimen length, L (in.)	
2. Specimen width, B (in.)	
3. Specimen height, H (in.)	
4. Mass of porcelain dish + dry sand (before use), M_1 (g)	
5. Mass of porcelain dish + dry sand (after use), M_2 (g)	
6. Dry unit weight of specimen, γ_d (lb/ft^3) $= \dfrac{M_1 - M_2}{LBH\,(\text{in.}^3)} \times 3.808$	
7. Specific gravity of soil solids, G_s	
8. Void ratio, $e = \dfrac{G_s \gamma_w}{\gamma_d} - 1$ (*Note:* $\gamma_w = 62.4$ lb/ft^3; γ_d is in lb/ft^3)	

Direct Shear Test on Sand

Stress and Displacement Calculation

Description of soil _____ Sample no. _____

Location _____

Normal load N _____ lb   Void ratio e _____

Tested by _____ Date _____

Normal Stress σ' (lb/in.^{2}) (1)	Horizontal Displacement (in.) (2)	Vertical Displacement* (in.) (3)	No. of div. in Proving Ring Dial Gauge (4)	Proving Ring Calibration Factor (lb/div.) (5)	Shear Force S (lb) (6)	Shear Stress τ (lb/in.^{2}) (7)

* Plus (+) sign means expansion.

Direct Shear Test on Sand

Stress and Displacement Calculation

Description of soil _____ Sample no. _____

Location _____

Normal load N _____ lb   Void ratio e _____

Tested by _____ Date _____

Normal Stress σ' (lb/in.^{2}) (1)	Horizontal Displacement (in.) (2)	Vertical Displacement* (in.) (3)	No. of div. in Proving Ring Dial Gauge (4)	Proving Ring Calibration Factor (lb/div.) (5)	Shear Force S (lb) (6)	Shear Stress τ (lb/in.^{2}) (7)

* Plus (+) sign means expansion.

Direct Shear Test on Sand

Stress and Displacement Calculation

Description of soil _____ Sample no. _____

Location _____

Normal load N _____ lb   Void ratio e _____

Tested by _____ Date _____

Normal Stress σ' (lb/in.^{2}) (1)	Horizontal Displacement (in.) (2)	Vertical Displacement* (in.) (3)	No. of div. in Proving Ring Dial Gauge (4)	Proving Ring Calibration Factor (lb/div.) (5)	Shear Force S (lb) (6)	Shear Stress τ (lb/in.^{2}) (7)

* Plus (+) sign means expansion.

Unconfined Compression Test

Description of soil _____ Specimen no. _____

Location _____

Moist mass of specimen _____ g   Moisture content _____ %

Length of specimen L _____ in.   Diameter of specimen _____ in.

Proving ring calibration factor: 1 div. = _____ lb   Area $A_0 = \frac{\Pi}{4}D^2 =$ _____ in.^{2}

Tested by _____ Date _____

Specimen Deformation ΔL (in.) (1)	Vertical Strain $\varepsilon = \frac{\Delta L}{L}$ (2)	Proving Ring Dial Reading (no. of small div.) (3)	Load P (Column 3 × calibration factor) (lb) (4)	Corrected Area $A_c = \frac{A_0}{1-\varepsilon}$ (in.^{2}) (5)	Stress $\sigma = \frac{\text{Column 4}}{\text{Column 5}}$ (lb/in.^{2}) (6)

Consolidation Test

Time versus Vertical Dial Reading

Description of soil _____

Location _____

Tested by _____ Date _____

Pressure on specimen _____ ton/ft^2 Pressure on specimen _____ ton/ft^2

Time after Load Application t (min)	$\sqrt{t}$ (min)$^{0.5}$	Vertical Dial Reading (in)	Time after Load Application t (min)	$\sqrt{t}$ (min)$^{0.5}$	Vertical Dial Reading (in.)

Consolidation Test
Time versus Vertical Dial Reading

Description of soil _____

Location _____

Tested by _____ Date _____

Pressure on specimen _____ ton/ft^2 Pressure on specimen _____ ton/ft^2

Time after Load Application t (min)	$\sqrt{t}$ (min)$^{0.5}$	Vertical Dial Reading (in)	Time after Load Application t (min)	$\sqrt{t}$ (min)$^{0.5}$	Vertical Dial Reading (in.)

Consolidation Test
Time versus Vertical Dial Reading

Description of soil _____

Location _____

Tested by _____ Date _____

Pressure on specimen _____ ton/ft^2 Pressure on specimen _____ ton/ft^2

Time after Load Application t (min)	$\sqrt{t}$ (min)$^{0.5}$	Vertical Dial Reading (in)	Time after Load Application t (min)	$\sqrt{t}$ (min)$^{0.5}$	Vertical Dial Reading (in.)

Consolidation Test
Pressure, Void Ratio, and Calculation of Coefficient of Consolidation

Description of soil _____ Location _____

Specimen diameter _____ in.   Initial specimen height $H_{t(i)}$ _____ in

Height of solids H_s _____ cm = _____ in.

Moisture content: Beginning of test _____ %   End of test _____ %

Mass of dry soil specimen _____ g G_s _____

Tested by _____   Date _____

Pressure p (ton/ft^2)	Final Dial Reading (in.)	Change in Specimen Height $\triangle H$ (in.)	Final Specimen Height $H_{t(f)}$ (in.)	Height of Void H_v (in.)	Final void ratio e	Average Height during Consolidation $H_{t(av)}$ (in.)	Fitting Time (s)		$C_v \times 10^3$ (in.^{2}/sec) from	
							t_{90}	t_{50}	t_{90}	t_{50}
(1)	(2)	(3)	(4)	(5)	(6)	(7)	(8)	(9)	(10)	(11)

Unconsolidated-Undrained Triaxial Test

Preliminary Data

Description of soil _____ Specimen no. _____

Location _____

Tested by _____ Date _____

Item	Quantity
1. Moist mass of specimen (end of test), M_1	
2. Dry mass of specimen, M_2	
3. Moisture content (end of test), $w\ (\%) = \dfrac{M_1 - M_2}{M_2} \times 100$	
4. Initial average length of specimen, L_0	
5. Initial average diameter of specimen, D_0	
6. Initial area, $A_0 = \dfrac{\pi}{4} D_0^2$	
7. Specific gravity of soil solids, G_s	
8. Final degree of saturation	
9. Cell confining pressure, σ_3	
10. Proving ring calibration factor	

Unconsolidated-Undrained Triaxial Test

Axial Stress–Strain Calculation

Specimen Deformation ΔL (in.)	Vertical Strain $\varepsilon = \frac{\Delta L}{L_0}$	Proving Ring Dial Reading (no. of small div.)	Piston Load P (Column 3 × calibration factor) (lb)	Corrected Area $A = \frac{A_0}{1-\varepsilon}$ (in.^{2})	Deviatory Stress $\Delta \sigma = \frac{P}{A}$ (lb/in.^{2})
(1)	(2)	(3)	(4)	(5)	(6)

Consolidated-Undrained Triaxial Test

Preliminary Data

Description of soil _____ Specimen no. _____

Location _____

Tested by _____ Date _____

Beginning of Test	
1. Moist unit weight of specimen (beginning of test)	
2. Moisture content (beginning of test)	
3. Initial length of specimen, L_0	
4. Initial diameter of specimen, D_0	
5. Initial area of specimen, $A_0 = \dfrac{\pi}{4}D_0^2$	
6. Initial volume of specimen, $V_0 = A_0 L_0$	
After Consolidation of Saturated Specimen	
7. Cell consolidation pressure, σ_3	
8. Net drainage from specimen during consolidation, ΔV	
9. Volume of specimen after consolidation, $V_0 - \Delta V = V_c$	
10. Length of specimen after consolidation, $L_c = L_0 \left(\dfrac{V_c}{V_0}\right)^{1/3}$	
11. Area of specimen after consolidation, $A_c = A_0 \left(\dfrac{V_c}{V_0}\right)^{2/3}$	

Consolidated-Undrained Triaxial Test

Axial Stress–Strain Calculation

Proving ring calibration factor _____

Specimen Deformation ΔL (cm) (1)	Vertical Strain $\varepsilon = \frac{\Delta L}{L_c}$ (2)	Proving Ring Dial Reading (no. of small div.) (3)	Piston Load P (N) (4)	Corrected Area $A = \frac{A_c}{1-\varepsilon}$ (cm^2) (5)	Deviatory Stress $\Delta\sigma = \frac{P}{A}$ (kN/m^2) (6)	Excess Pore-Water Pressure Δu (kN/m^2) (7)	$\bar{A} = \frac{\Delta u}{\Delta\sigma}$ (8)

Consolidated-Undrained Triaxial Test

Preliminary Data

Description of soil _____ Specimen no. _____

Location _____

Tested by _____ Date _____

Beginning of Test	
1. Moist unit weight of specimen (beginning of test)	
2. Moisture content (beginning of test)	
3. Initial length of specimen, L_0	
4. Initial diameter of specimen, D_0	
5. Initial area of specimen, $A_0 = \dfrac{\pi}{4}D_0^2$	
6. Initial volume of specimen, $V_0 = A_0 L_0$	
After Consolidation of Saturated Specimen	
7. Cell consolidation pressure, σ_3	
8. Net drainage from specimen during consolidation, ΔV	
9. Volume of specimen after consolidation, $V_0 - \Delta V = V_c$	
10. Length of specimen after consolidation, $L_c = L_0 \left(\dfrac{V_c}{V_0}\right)^{1/3}$	
11. Area of specimen after consolidation, $A_c = A_0 \left(\dfrac{V_c}{V_0}\right)^{2/3}$	

Consolidated-Undrained Triaxial Test

Axial Stress–Strain Calculation

Proving ring calibration factor _____

Specimen Deformation ΔL (cm)	Vertical Strain $\varepsilon = \frac{\Delta L}{L_c}$	Proving Ring Dial Reading (no. of small div.)	Piston Load P (N)	Corrected Area $A = \frac{A_c}{1-\varepsilon}$ (cm²)	Deviatory Stress $\Delta\sigma = \frac{P}{A}$ (kN/m²)	Excess Pore-Water Pressure Δu (kN/m²)	$\bar{A} = \frac{\Delta u}{\Delta\sigma}$
(1)	(2)	(3)	(4)	(5)	(6)	(7)	(8)

APPENDIX E

Graph Paper

There are ten sheets of 5-cycle, five sheets of 3-cycle, five sheets of 1-cycle semilog graph paper, and ten sheets of linear graph paper included in this appendix. You may copy more pages if needed.

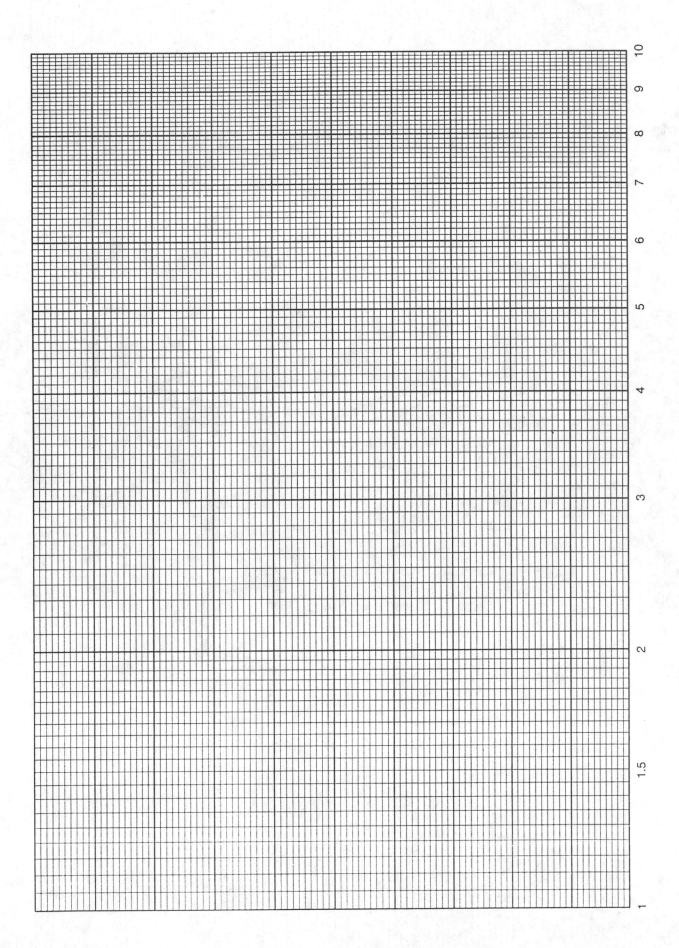

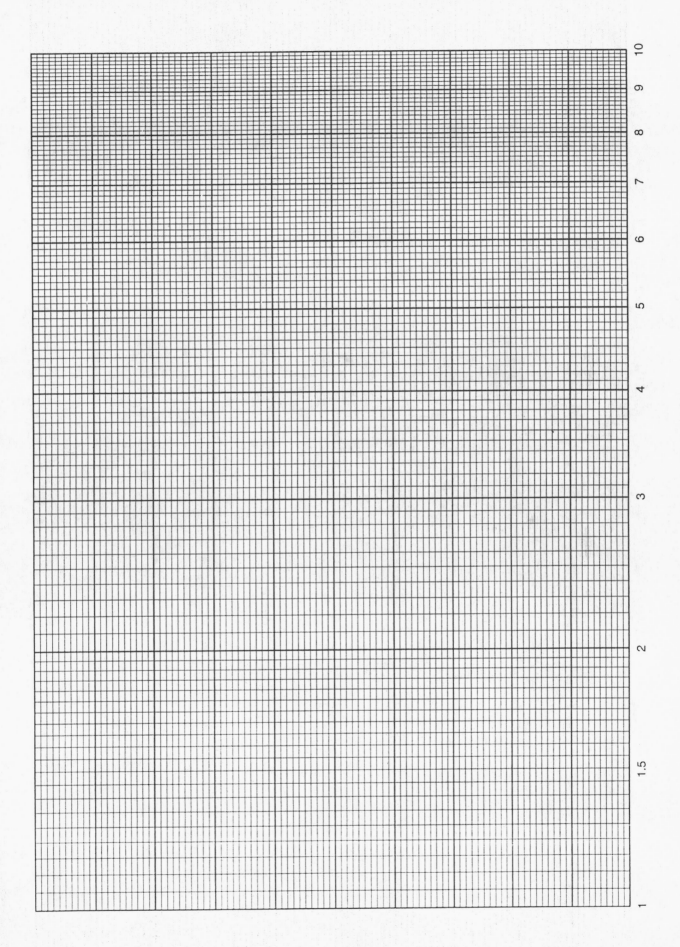

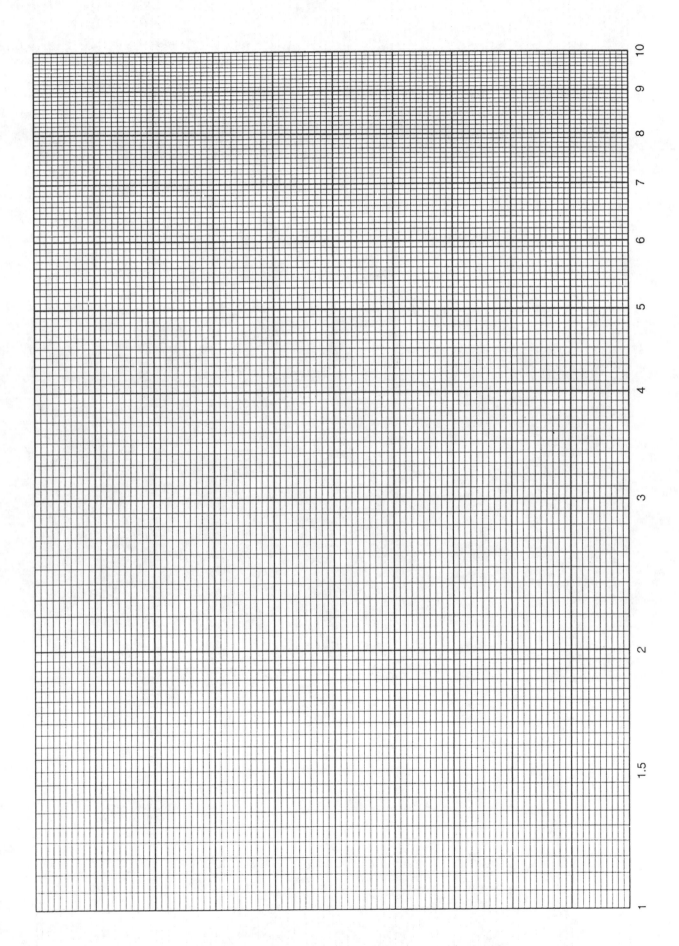

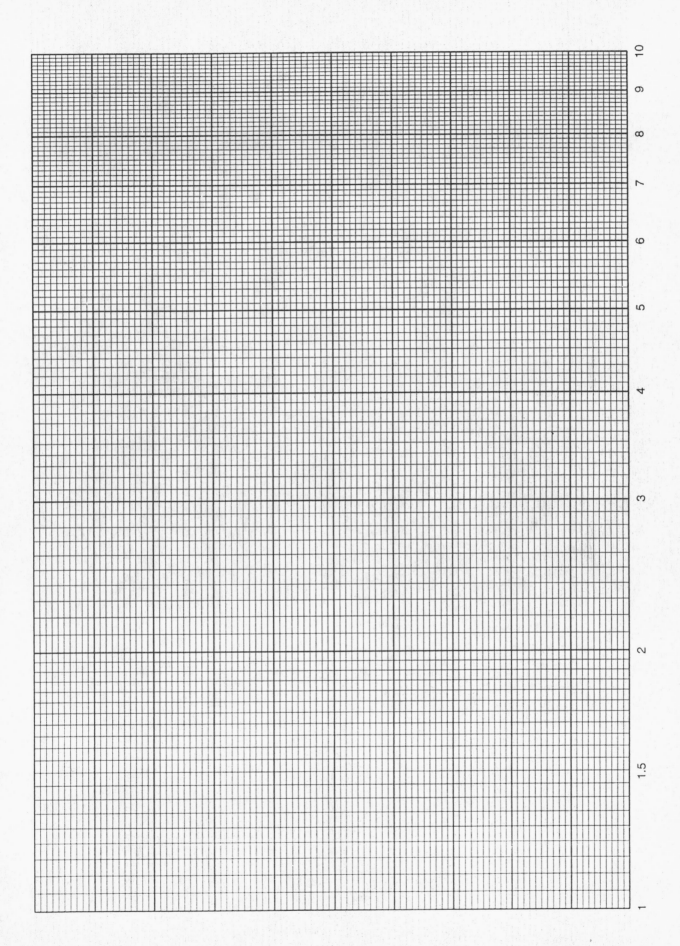

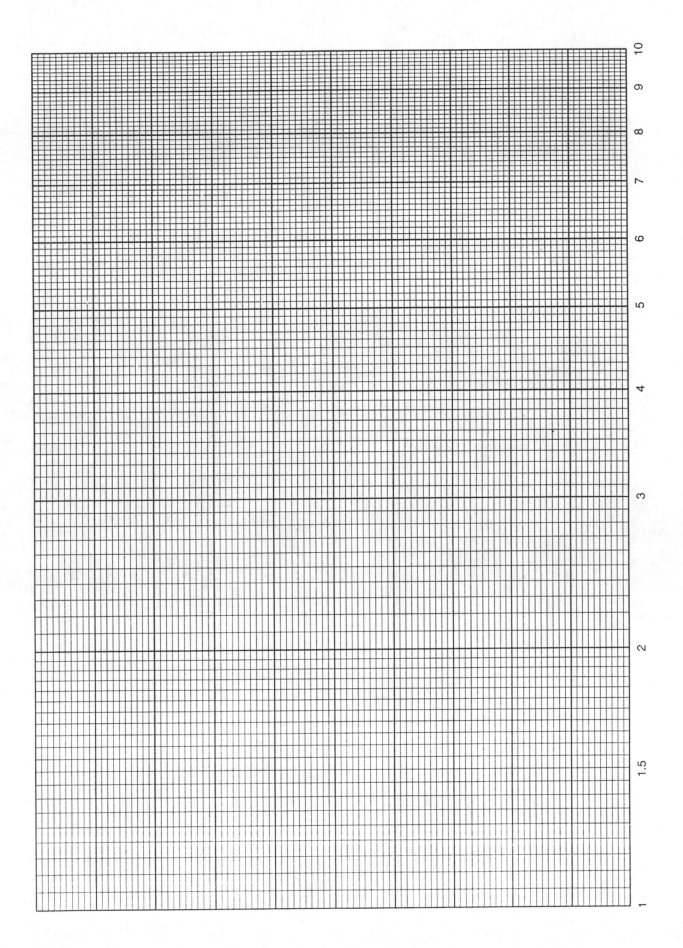

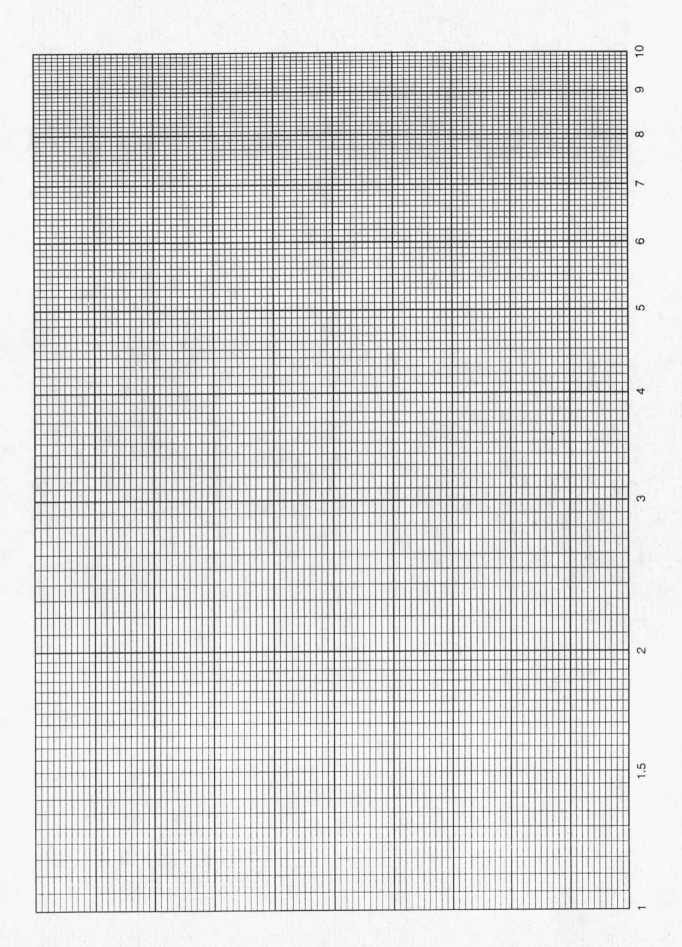

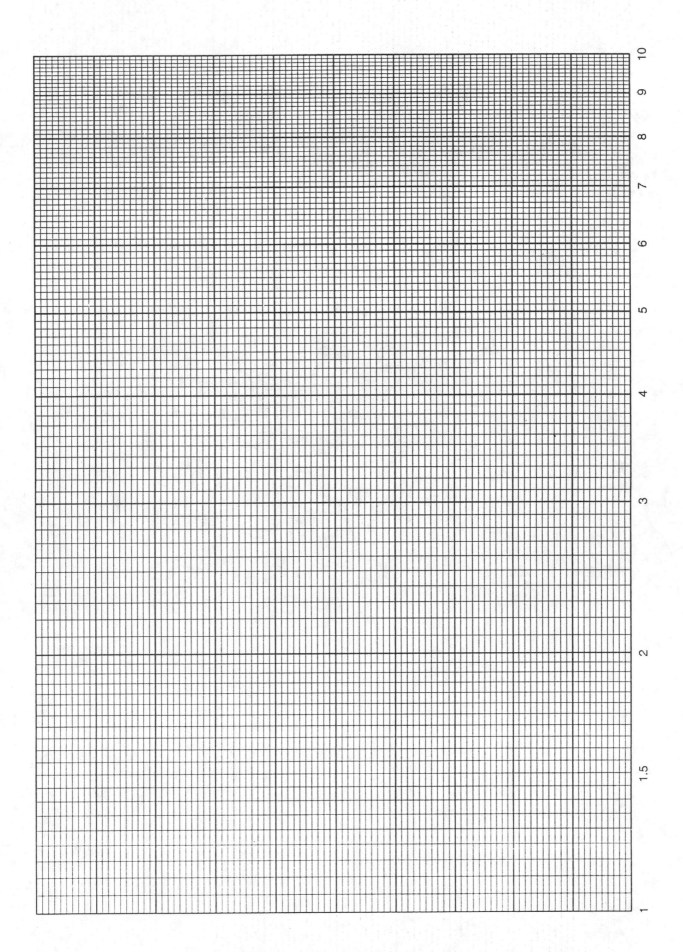

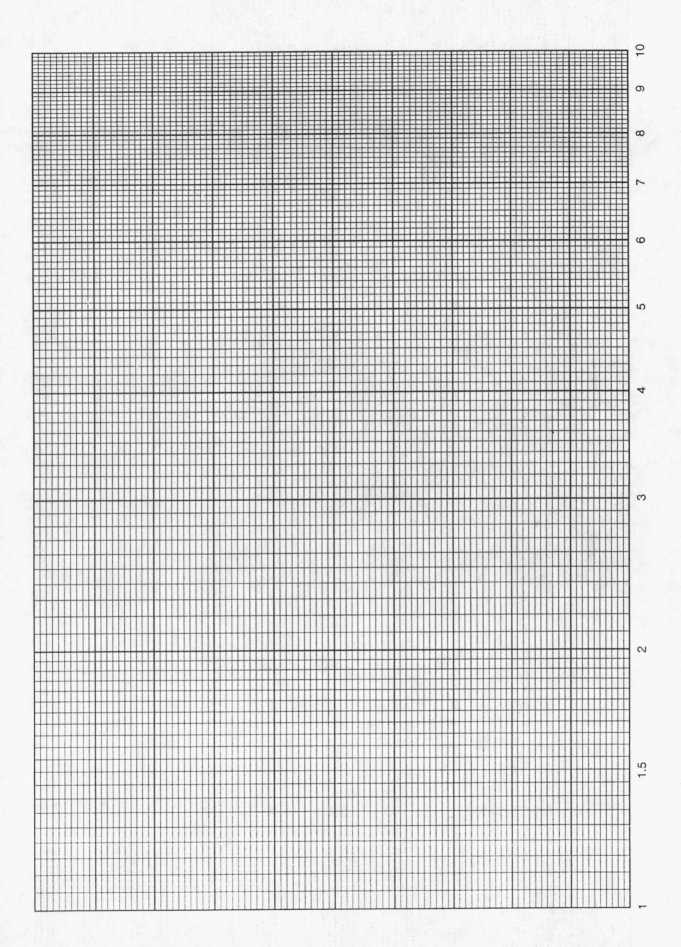

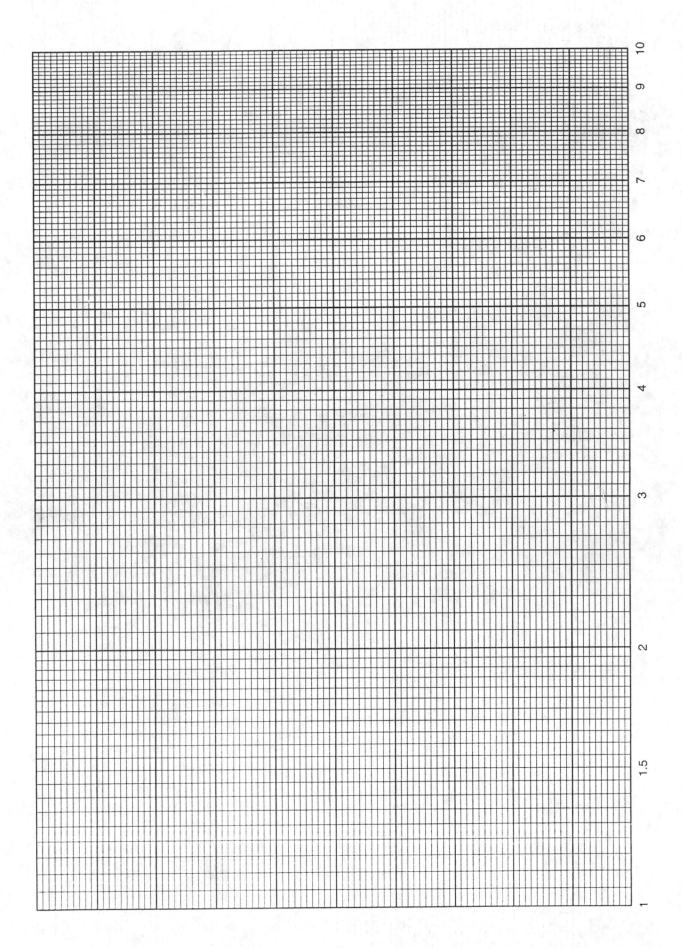

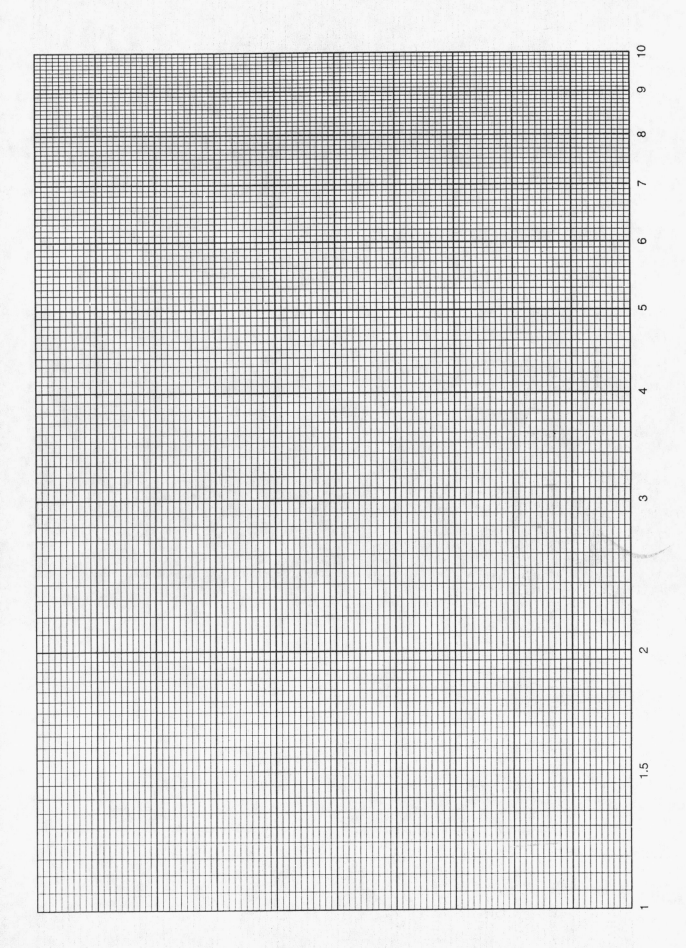

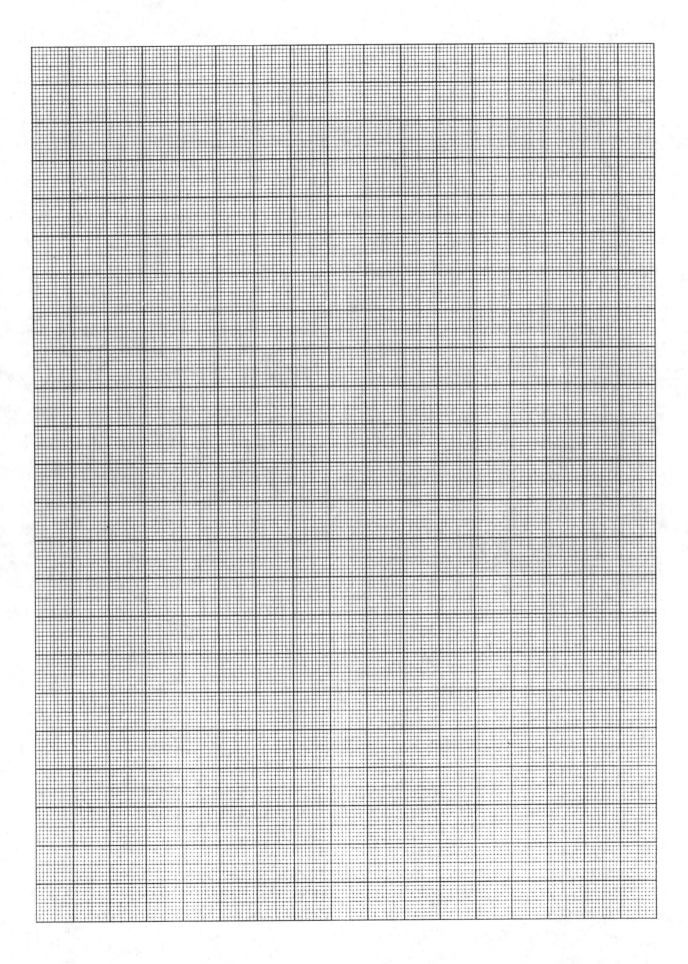

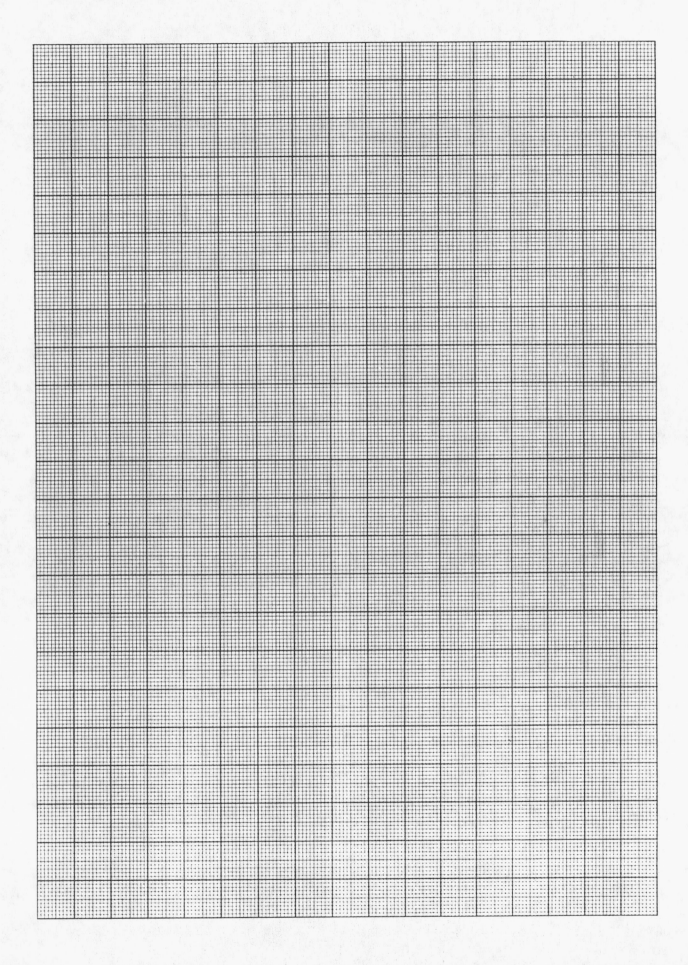

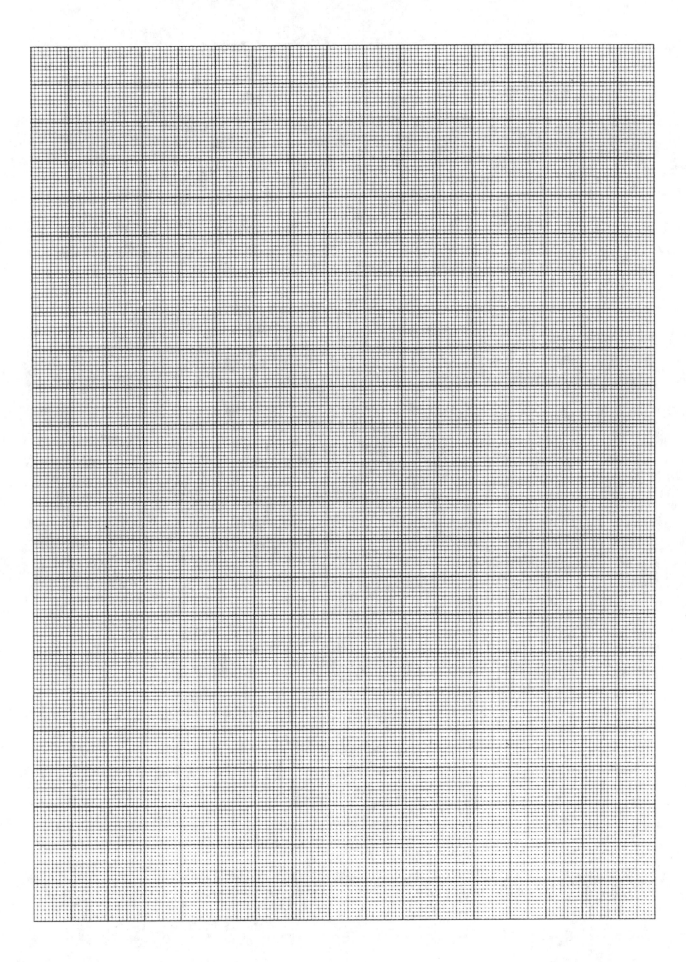

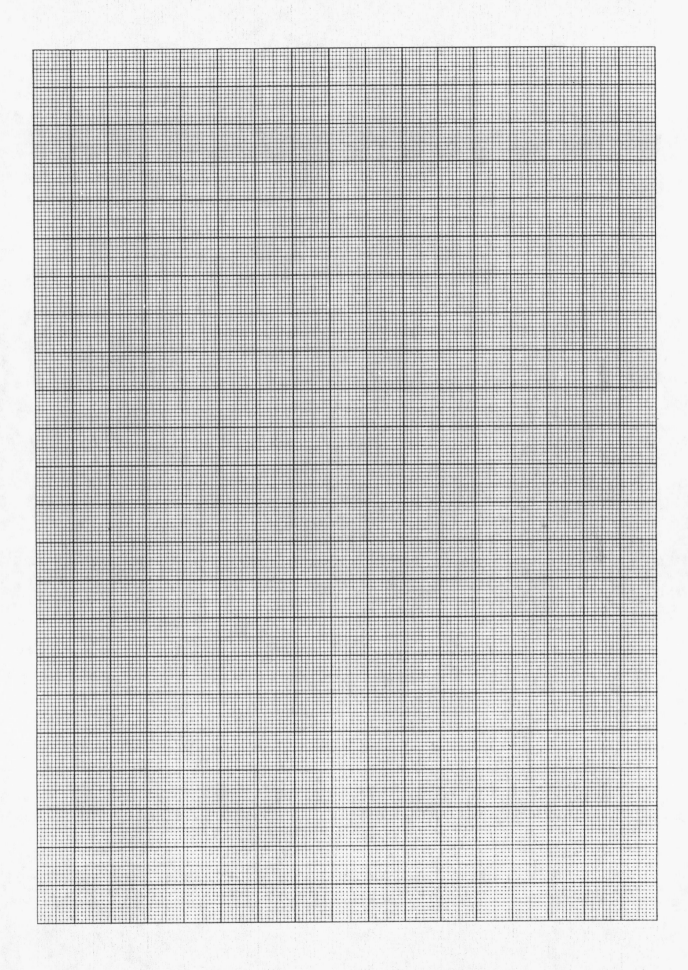

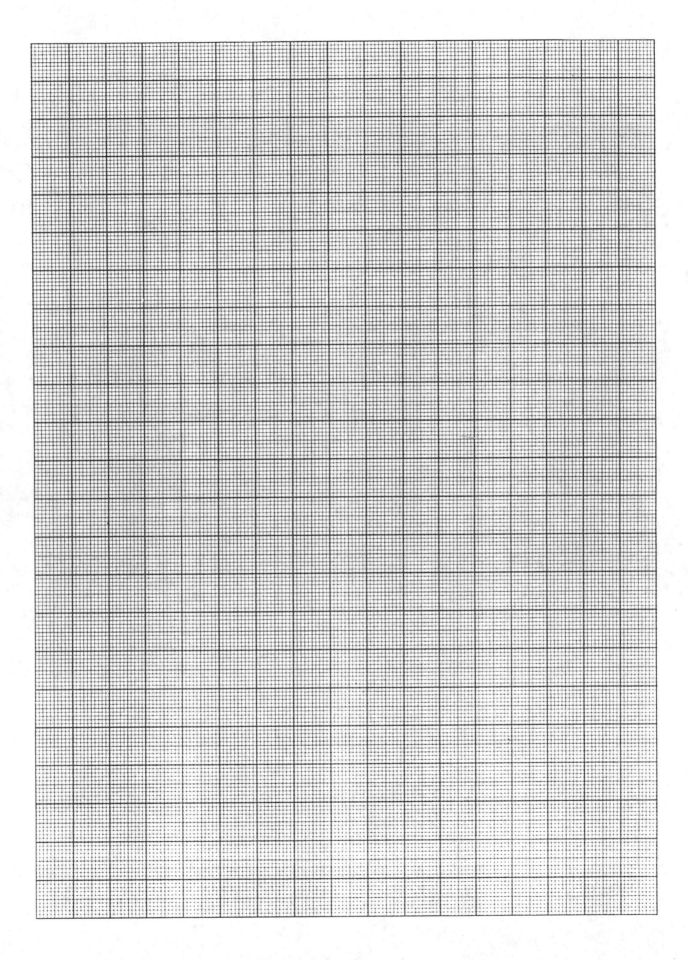

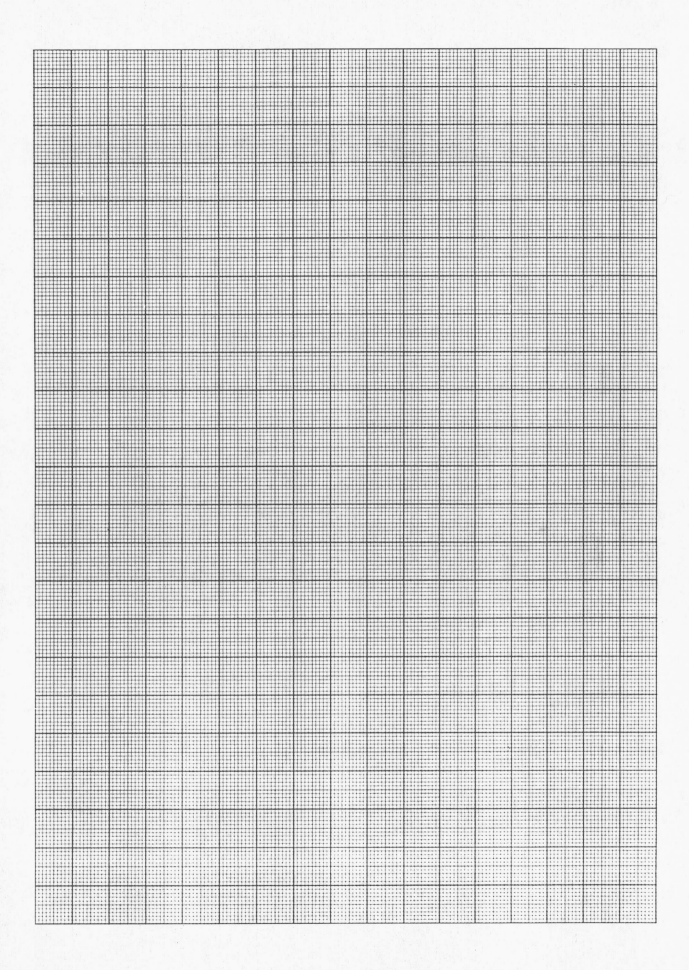

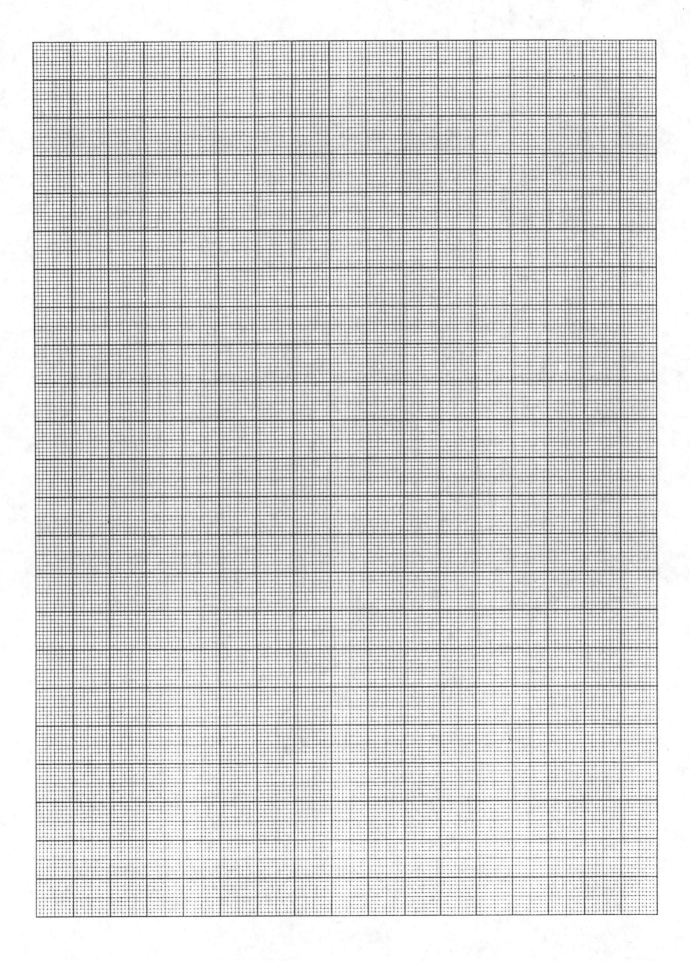

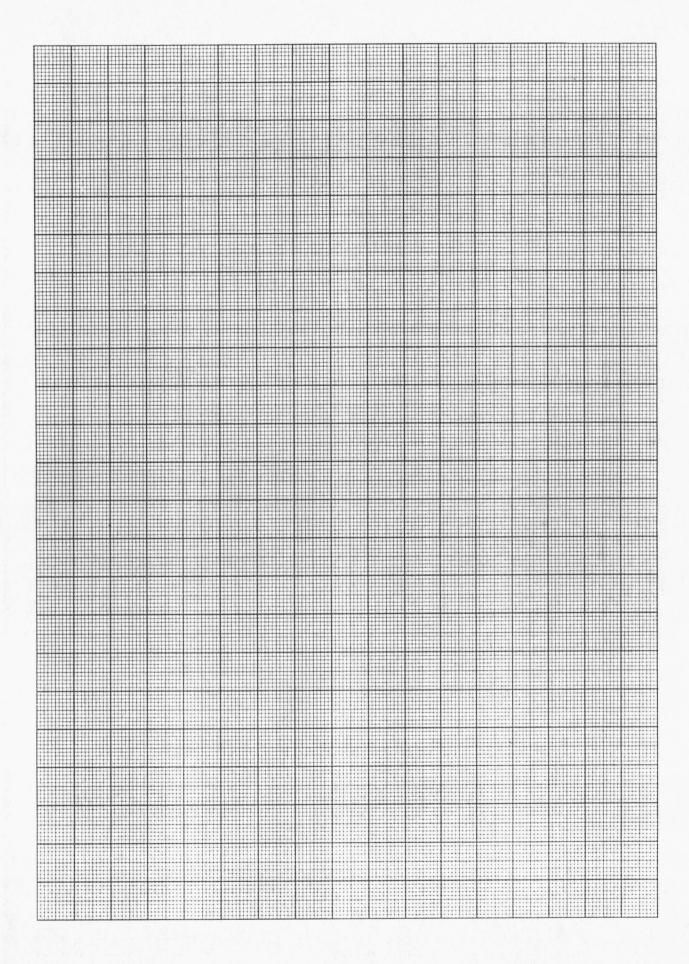

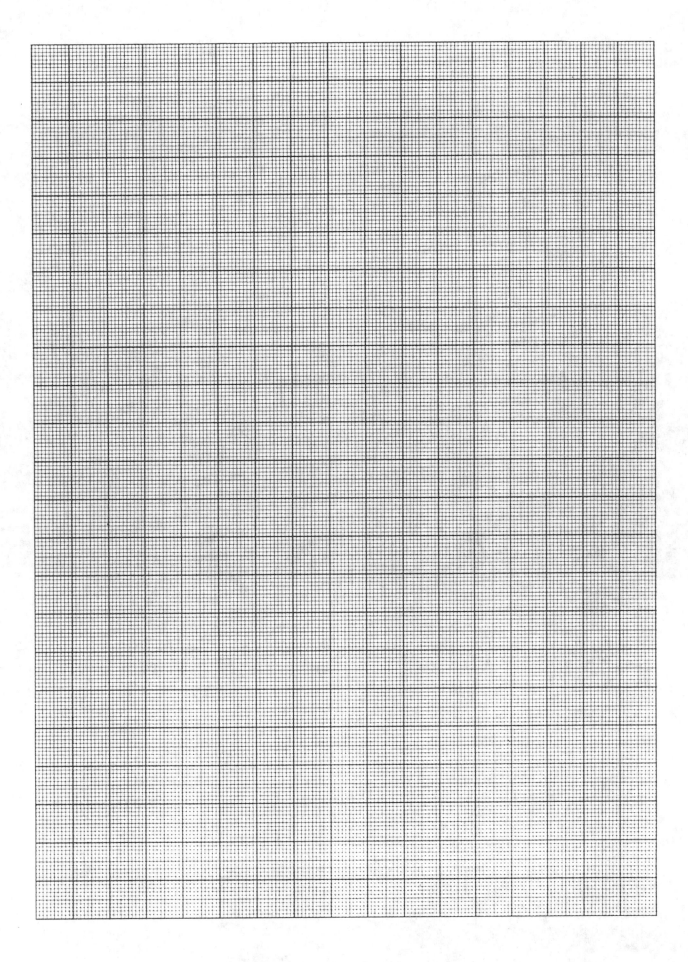

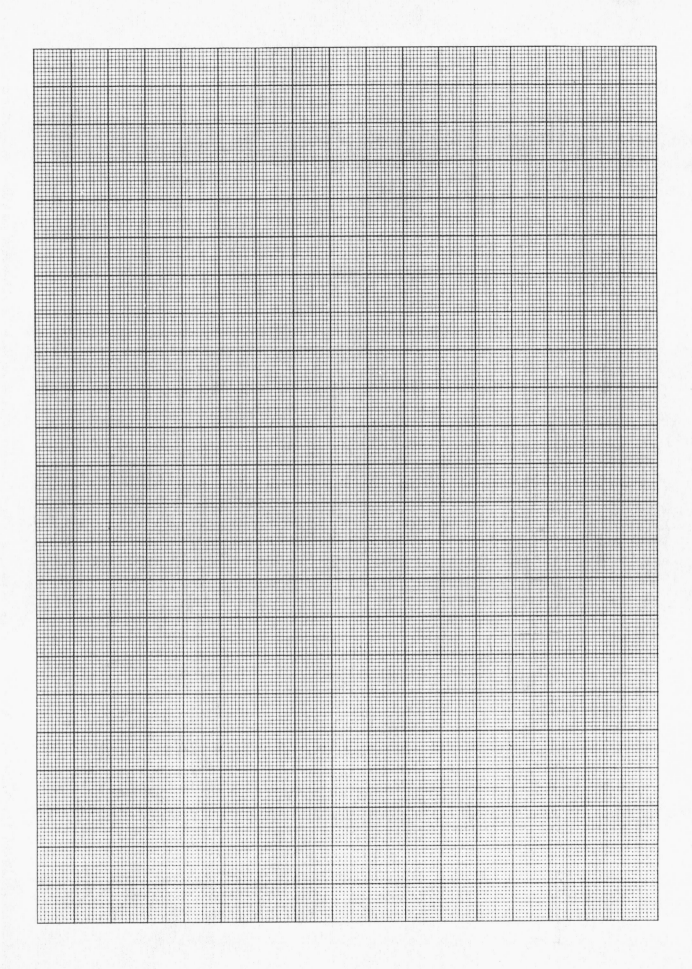

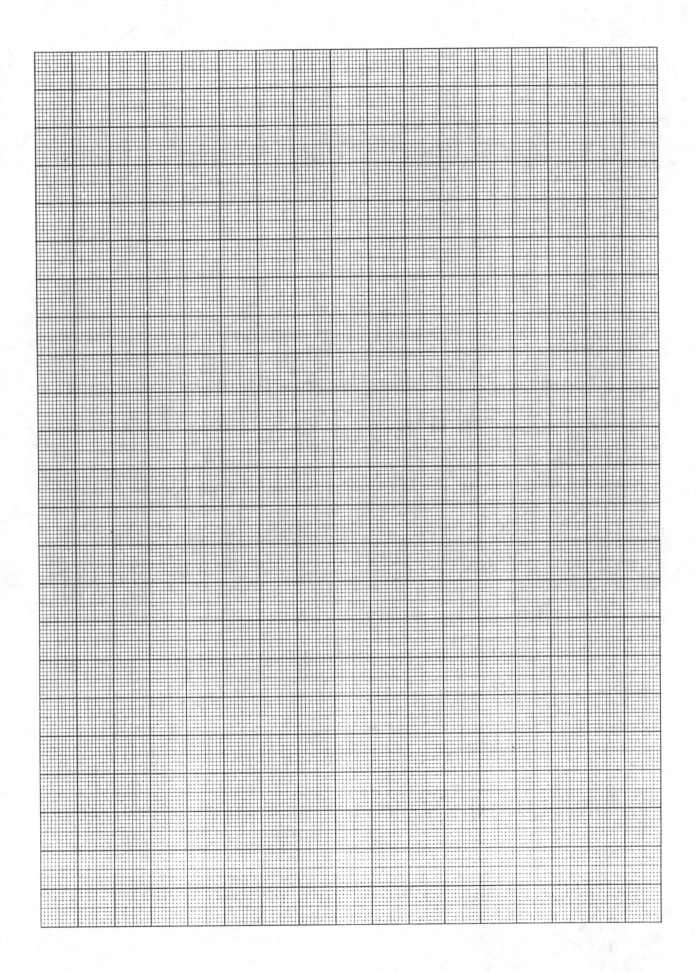

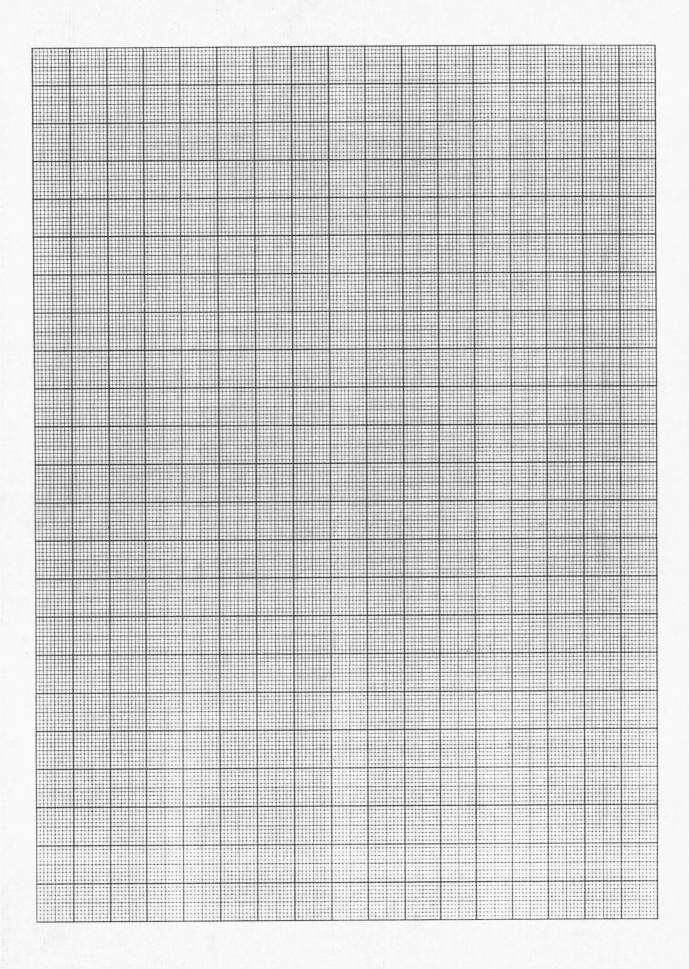

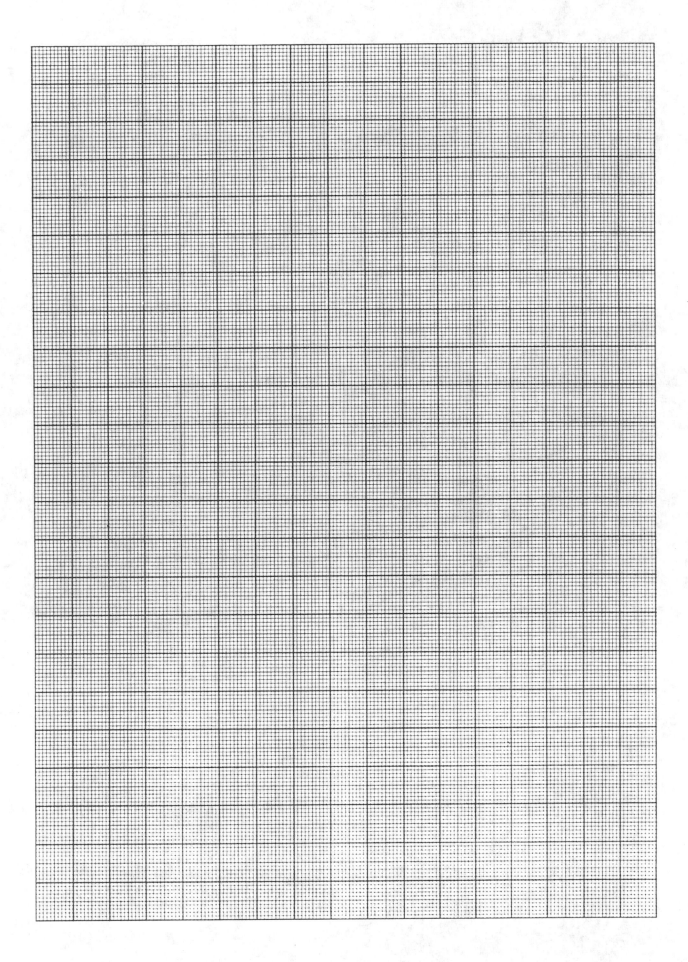

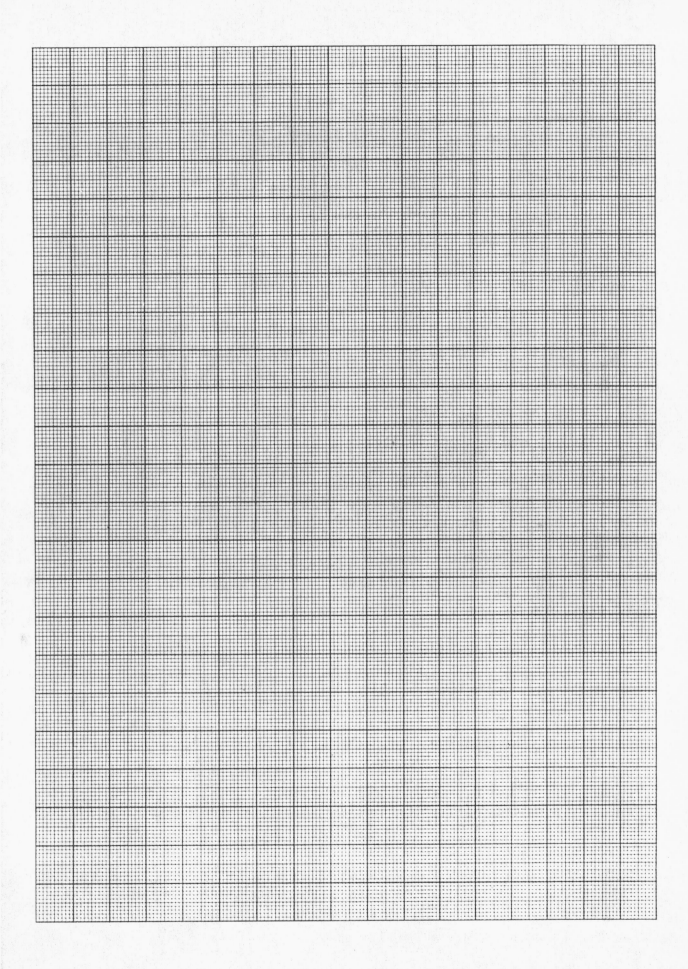

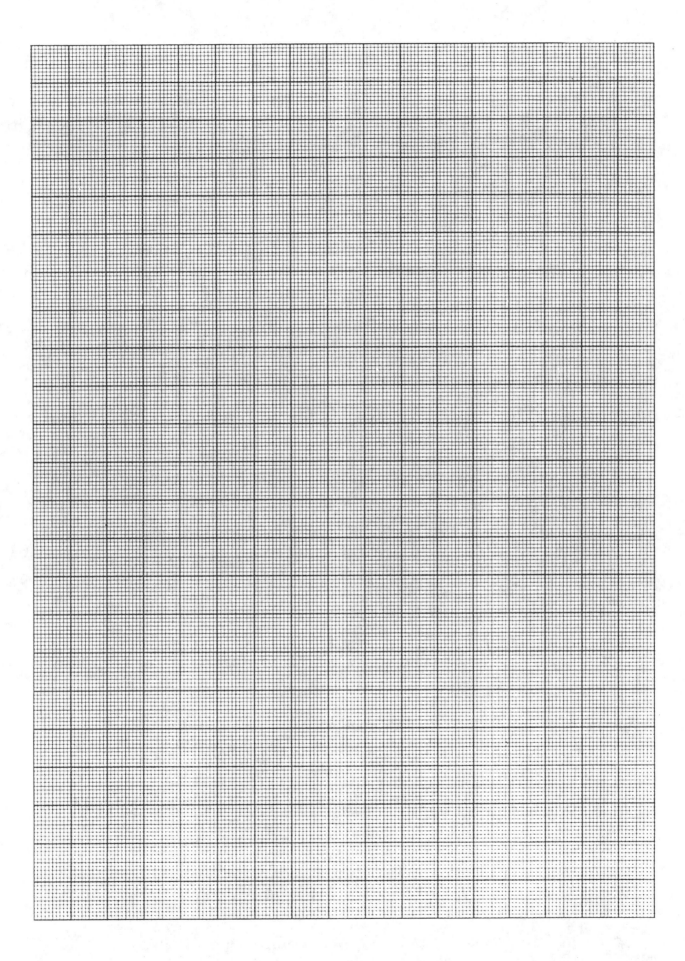

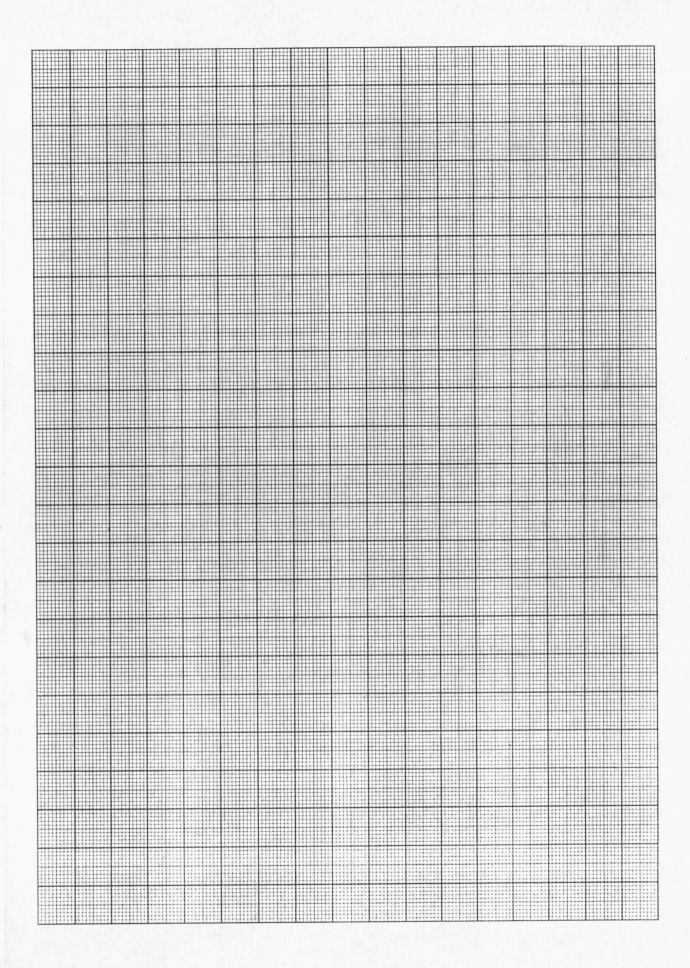

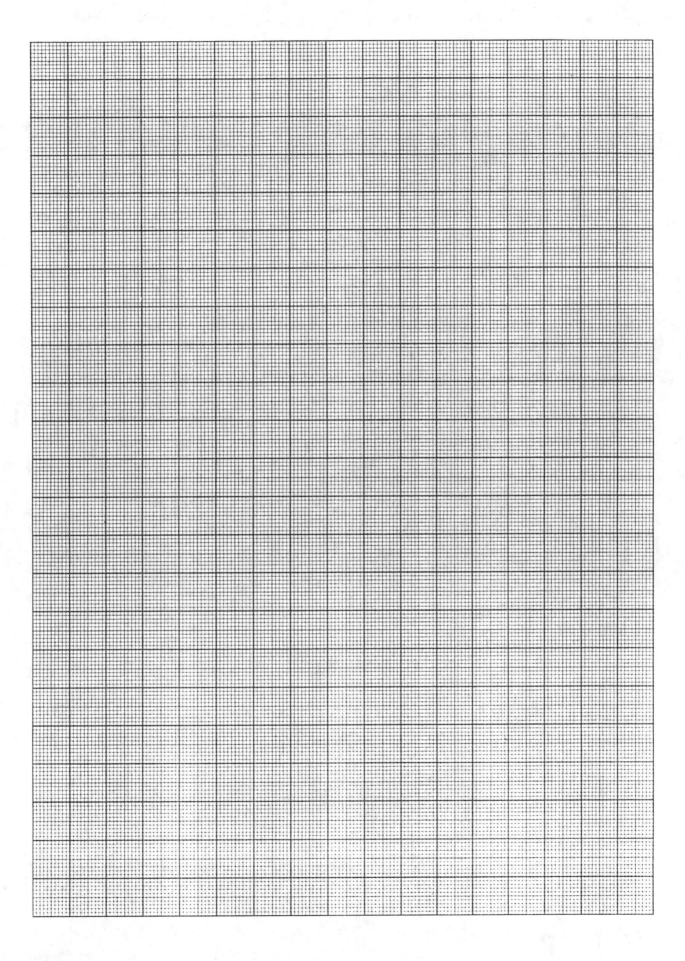

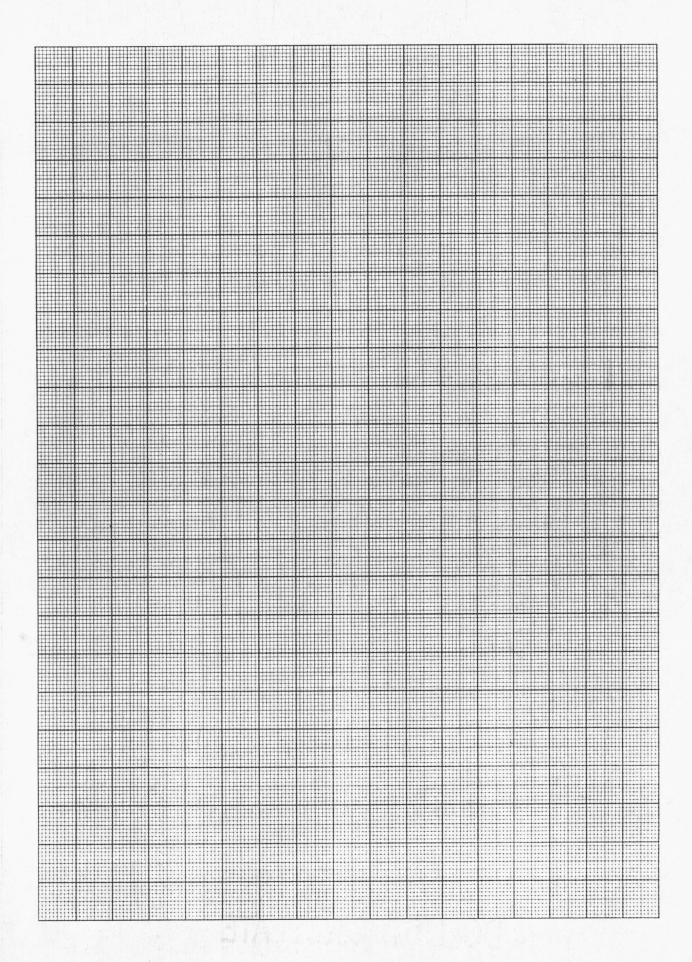